ENGINEERING SCIENCE II

Macmillan Technician Series

P. Astley, *Engineering Drawing and Design II*

P. J. Avard and J. Cross, *Workshop Processes and Materials I*

G. D. Bishop, *Electronics II*

G. D. Bishop, *Electronics III*

John Elliott, *Building Science and Materials II/III*

J. Ellis and N. Riches, *Safety and Laboratory Practice*

P. R. Lancaster and D. Mitchell, *Mechanical Science III*

R. Lewis, *Physical Science I*

Noel M. Morris, *Electrical Principles II*

Noel M. Morris, *Electrical Principles III*

Engineering Science II

D. E. Hewitt

C. Eng., M. I. Mech. E.,

Lecturer in Mechanical Engineering
Department of Mechanical and Production Engineering,
Loughborough Technical College

First published 1978 by
THE MACMILLAN PRESS LTD
London and Basingstoke
Associated companies in Delhi Dublin
Hong Kong Johannesburg Lagos Melbourne
New York Singapore and Tokyo

Typeset in 10/12 *Times*
Printed and bound in Great Britain by
A. Wheaton & Co. Ltd., Exeter

British Library Cataloguing in Publication Data

Hewitt, Derrick Edward
Engineering Science II.—(Macmillan technician series).
1. Engineering
I. Title II. Series
620 TA145

ISBN 0-333-21391-2

Contents

Section	Page	U75/036	U76/053	U76/061	U76/062	U76/063
Foreword	ix					
Preface	xi					
1. Statics—Materials	1					
1.1 Stress	1	√	√	√	√	√
1.2 Strain	2	√	√	√	√	√
1.3 Hooke's Law and the Modulus of Elasticity	3	√	√	√	√	√
1.4 Elasticity and Plasticity	5	√	√	√	√	√
1.5 The Tensile Test	6	√	√		√	√
1.6 Properties of Materials and Stress–Strain Curves	11	√	√		√	√
1.7 Introduction to Plastics Materials	12				√	√
1.8 Shear Stress and Shear Strain—Modulus of Rigidity	13			√		
1.9 Factor of Safety and Working Stresses	14			√		
Exercises	16					
2. Statics—Structures	18					
2.1 Force as a Vector	18	√	√	√	√	√
2.2 Equilibrium—The Triangle and Polygon of Forces	20	√	√	√	√	
2.3 Bow's Notation	21	√		√	√	
2.4 Conditions of Equilibrium	22	√	√	√	√	√
2.5 The Turning Effect of a Force	24	√	√	√	√	√
2.6 Beam Reactions	25	√	√	√	√	√
2.7 Resultant Forces in Frameworks	26	√		√	√	
2.8 Graphical Determination of Forces in Simply Supported Frameworks Subject to Vertical Loads	27	√		√	√	
2.9 Graphical Determination of Forces in Simply Supported Frameworks Subject to Vertical and Non-vertical Loads	28				√	
2.10 Graphical Determination of Forces in Cantilever Frameworks	30				√	
Exercises	31					
3. Dry Sliding Friction	39					
3.1 The Nature of Dry Sliding Friction	39	√	√	√		

Section	Page	U75/036	U76/053	U76/061	U76/062	U76/063
3.2 Angle of Friction (Angle of Repose)	42	√	√	√		
3.3 Full-film and Boundary Lubrication	43			√		
3.4 Friction Losses in Bearings	44	√	√	√		
Exercises	46					
4. Simple Machines	48					
4.1 Simple Machines and Levers	48	√	√	√		
4.2 Mechanical Advantage and Velocity or Distance Ratio	49	√	√	√		
4.3 Velocity Ratio or Distance Ratio of Simple Machines	50	√	√	√		
4.4 Mechanical Efficiency of Simple Machines	55	√	√	√		
4.5 Limiting Efficiency—the Law of a Simple Machine	57	√	√	√		
4.6 Overhauling in a Simple Machine—Ideal Machines	59			√		
Exercises	61					
5. Velocity	62					
5.1 Constant-velocity and Uniformly Accelerated Motion	62			√	√	
5.2 Distance–Time, Velocity–Time and Acceleration–Time Graphs for Uniformly Accelerated Motion	64			√	√	
5.3 Uniformly Accelerated Motion Due to Gravity	66			√	√	
5.4 Velocity as a Vector	67	√	√			
5.5 Relative Velocity Applied to Mechanisms	69	√	√			
5.6 Circular Motion and Angular Velocity	70	√	√	√	√	
Exercises	73					
6. Dynamics—Newton's Laws of Motion	76					
6.1 Newton's First Law of Motion	76	√	√	√	√	√
6.2 Newton's Second Law of Motion	77	√	√			
6.3 Newton's Third Law of Motion	79	√	√			
6.4 Forces Required to Produce and Maintain Motion	79	√			√	
Exercises	83					
7. Dynamics—Work and Energy	84					
7.1 Work and Energy	84	√			√	

Section	Page	U75/036	U76/053	U76/061	U76/062	U76/063
7.2 Work done by a Constant Force	85	✓			✓	
7.3 Work done by a Varying Force	85	✓				
7.4 Work done by an Inclined Force	86	✓			✓	
7.5 Mechanical Energy	87	✓		✓	✓	
7.6 Potential Energy	87	✓		✓	✓	
7.7 Kinetic Energy	87	✓		✓	✓	
7.8 Conservation of Energy	89	✓		✓	✓	✓
7.9 Power—the Rate of Doing Work	91	✓			✓	✓
Exercises	94					
8. Heat and Its Applications	95					
8.1 Heat and Energy	95			✓	✓	✓
8.2 Effects of Heat	95	✓	✓	✓	✓	✓
8.3 Sources of Heat	95			✓		
8.4 Heat Transfer	95	✓	✓			
8.5 Temperature	97	✓	✓		✓	✓
8.6 Calorimetry	98	✓	✓		✓	✓
8.7 Change of State	100	✓	✓		✓	✓
8.8 Change of Size	102	✓	✓	✓	✓	✓
8.9 Properties of Gases	104				✓	
Exercises	107					
9. Electricity and Electrical Measurement	109					
9.1 The Nature of Electric Charge	109	✓	✓		✓	✓
9.2 The Electric Current	110	✓	✓	✓	✓	✓
9.3 The Electrical Circuit	111	✓	✓	✓	✓	✓
9.4 Factors Determining Resistance	118			✓		
9.5 Electric Power	121					✓
9.6 Electrical Measurements	122	✓	✓	✓		✓
9.7 Measurement in A.C. Systems	128	✓	✓		✓	✓
9.8 Electrostatics	130				✓	
9.9 Electrochemistry	134					✓
9.10 Safety Precautions for Electrical Equipment	140	✓	✓	✓	✓	✓
Exercises	142					
10. Magnetism and Electromagnetism	144					
10.1 Evidence of a Magnetic Effect	144	✓	✓	✓		

	Section	Page	U75/036	U76/053	U76/061	U76/062	U76/063
10.2	Shape and Distribution of Magnetic Fields	145			✓		
10.3	The Concept of Magnetic Flux and Field Strength	147	✓	✓	✓	✓	
10.4	The Direct-current Motor	151	✓	✓	✓		
10.5	Electromagnetic Induction	155	✓	✓	✓	✓	
10.6	Machines for Generating Electric Current	158	✓	✓		✓	
10.7	Flux Change as an Alternative to Flux Cutting	160	✓	✓	✓	✓	
10.8	Self-inductance and Mutual Inductance	163	✓	✓		✓	
10.9	Production of Magnetism in Magnetic Materials	165			✓	✓	
Exercises		168					

Foreword

This book is written for one of the many technician courses now being run at technical colleges in accordance with the requirements of the **Technician Education Council** (TEC). This Council was established in March 1973 as a result of the recommendation of the Government's Haslegrave Committee on Technical Courses and Examinations, which reported in 1969. TEC's functions were to rationalise existing technician courses, including the City and Guilds of London Institute (C.G.L.I.) Technician courses and the Ordinary and Higher National Certificate courses (O.N.C. and H.N.C.), and provide a system of technical education which satisfied the requirements of 'industry' and 'students' but that could be operated economically and efficiently.

Four qualifications are awarded by TEC, namely the Certificate, Higher Certificate, Diploma and Higher Diploma. The **Certificate** award is comparable with the O.N.C. or with the third year of the C.G.L.I. Technician course, whereas the **Higher Certificate** is comparable with the H.N.C. or the C.G.L.I. Part III Certificate. The **Diploma** is comparable with the O.N.D. in Engineering or Technology, the **Higher Diploma** with the H.N.D. Students study on a part-time or block-release basis for the Certificate and Higher Certificate, whereas the Diploma courses are intended for full-time study. Evening study is possible but not recommended by TEC. The Certificate course consists of fifteen Units and is intended to be studied over a period of three years by students, mainly straight from school, who have three or more C.S.E. Grade III passes or equivalent in appropriate subjects such as mathematics, English and science. The Higher Certificate course consists of a further ten Units, for two years of part-time study, the total time allocation being 900 hours of study for the Certificate and 600 hours for the Higher Certificate. The Diploma requires about 2000 hours of study over two years, the Higher Diploma a further 1500 hours of study for a further two years.

Each student is entered on to a **Programme** of study on entry to the course; this programme leads to the award of a Technician Certificate, the title of which reflects the area of engineering or science chosen by the student, such as the Telecommunications Certificate or the Mechanical Engineering Certificate. TEC have created three main **Sectors** of responsibility

Sector A responsible for General, Electrical and Mechanical Engineering
Sector B responsible for Building, Mining and Construction Engineering
Sector C responsible for the Sciences, Agriculture, Catering, Graphics and Textiles.

Each Sector is divided into Programme committees, which are responsible for the specialist subjects or programmes, such as A1 for General Engineering, A2 for Electronics and Telecommunications Engineering, A3 for Electrical Engineering, etc. Colleges have considerable control over the content of their intended programmes, since they can choose the Units for their programmes to suit the requirements of local industry, college resources or student needs. These Units can be written entirely by the college, thereafter called a college-devised Unit, or can be supplied as a Standard Unit by one of the Programme committees of TEC. **Assessment** of every Unit is carried out by the college and a pass in one Unit depends on the attainment gained by the student in his coursework, laboratory work and an end-of-Unit test. TEC moderate college assessment plans and their validation; external assessment by TEC will be introduced at a later stage.

The three-year Certificate course consists of fifteen Units at three Levels: I, II and III, with five Units normally studied per year. A typical programme might be as follows.

Year I	Mathematics I	Standard Unit	six Level I Units
	Science I	Standard Unit	
	Workshop Processes I	Standard Unit	
	Drawing I	Standard Unit	

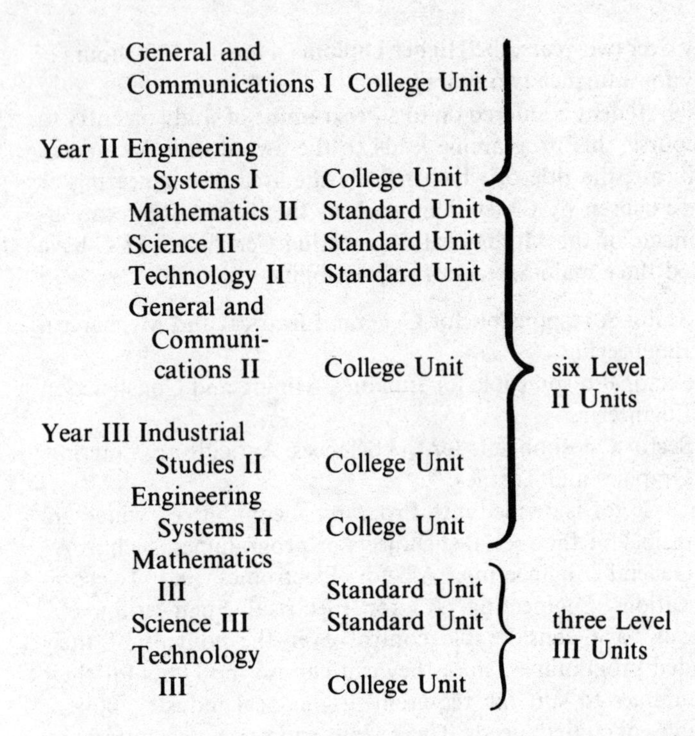

Year	Unit	Type	Grouping
	General and Communications I	College Unit	
Year II	Engineering Systems I	College Unit	
	Mathematics II	Standard Unit	six Level II Units
	Science II	Standard Unit	six Level II Units
	Technology II	Standard Unit	six Level II Units
	General and Communications II	College Unit	six Level II Units
Year III	Industrial Studies II	College Unit	six Level II Units
	Engineering Systems II	College Unit	six Level II Units
	Mathematics III	Standard Unit	three Level III Units
	Science III	Standard Unit	three Level III Units
	Technology III	College Unit	three Level III Units

Entry to each Level I or Level II Unit will carry a prerequisite qualification such as C.S.E. Grade III for Level I or O-level for Level II; certain Craft qualifications will allow students to enter Level II direct, one or two Level I Units being studied as 'trailing' Units in the first year. The study of five Units in one college year results in the allocation of about two hours per week per Unit, and since more subjects are often to be studied than for the comparable City and Guilds course, the treatment of many subjects is more general, with greater emphasis on an **understanding** of subject topics rather than their application. Every syllabus to every Unit is far more detailed than the comparable O.N.C. or C.G.L.I. syllabus, presentation in **Learning Objective** form being requested by TEC. For this reason a syllabus, such as that followed by this book, might at first sight seem very long, but analysis of the syllabus will show that 'in-depth' treatment is not necessary—objectives such as '. . . **states** Ohm's law . . . ' or ' . . . **lists** the different types of telephone receiver . . . ' clearly do **not** require an understanding of the derivation of the Ohm's law equation or the operation of several telephone receivers.

The treatment of each topic is carried to the depth suggested by TEC and in a similar way the **length** of the Unit (sixty hours of study for a full Unit), **prerequisite qualifications, credits for alternative qualifications** and **aims of the Unit** have been taken into account by the author.

Preface

The aim of this book is concisely to cover sufficient engineering science material at Level II to meet the needs of any student following a Technician Education Council programme of studies within the fields of Sector A1—General Engineering and Sector A5—Mechanical and Production Engineering.

The standard Units provided by TEC to meet the needs of the outline programmes in these areas are

TEC U76/053 Engineering Science—aiming to give a basic mechanical and electrical engineering science background for engineering manufacturing technology; taken by *all* students in the mechanical and production engineering field.

TEC U76/061 Engineering Science (1) II—a half Unit aiming to provide a basic background of engineering science for *all* technician students in the general engineering field.

TEC U76/062 Engineering Science (2) A II—an additional half Unit to develop the engineering science background of the more analytical student following a Certificate programme in general engineering.

TEC U76/063 Engineering Science (2)B II—an additional half Unit to complement TEC U76/061 to give the more practical student following a Certificate programme in general engineering an engineering science background to his technology subjects.

The students taking these particular Units will include those just entering their further education studies, having obtained an O-level G.C.E. pass (Grade 3 or better) in physics or a suitable science, or, alternatively, having obtained a C.S.E. Grade 1 pass in similar subjects.

Other students will include those whose school-leaving attainments are less than those outlined above, but who have successfully studied TEC Units TEC U75/004 Physical Science I or TEC U76/004 Engineering Science I. There will also be students transferring to a technician course having distinguished themselves on a craft course.

The material within the book is presented in logically arranged chapters, each preceded by its own general objective. Each chapter is sub-divided into concise sections, which any student may select to study to fulfil the specific objectives associated with his programme. Each sub-section is cross-referenced to the particular TEC Unit or half Unit to which it is relevant. Each chapter concludes with specific objectives and these, coupled with the student exercises, should help the student to monitor his own progress effectively and ensure that the specific objectives have been achieved.

There are numerous worked examples within the text, carefully graded in complexity, illustrating where possible the application of the principles involved to everyday and common engineering situations.

The text also includes brief biographical details of some of the scientists and engineers to whom we owe much of our current knowledge. Comment has also been included to emphasise the aspects of science and technology which have particular social and environmental implications. It is hoped that this will assist in the integration of general studies by providing introductory material for discussion on topics such as the history of technology, energy sources, pollution, etc., and will broaden the interests of the student and make him aware of the importance of the engineering industry in society and, in particular, the role of the 'technician' within the industry.

I wish to acknowledge the assistance and collaboration of my colleague, C. H. Salsbury, B.Sc., C.Eng., M.I. Gas E., M.Inst.F., in the preparation of the book, in particular for reading through the manuscript and making helpful comments, and for his contribution of chapters 8, 9 and 10 on heat, electricity and magnetism.

D. E. Hewitt

1 Statics—Materials

The object of this chapter is that the student shall appreciate the effects of forces on materials used in engineering and the need for care in design and materials selection.

1.1 STRESS

While force itself cannot be seen, its effects may be felt and seen. A force is defined not by what it is, but by what it does. The most obvious effect of a force is that motion may be produced or modified. Indeed, the definition of a force according to Newton's laws of motion (section 6.1) is that a force is that which changes or tends to change motion. That is, when a body is at rest, a force will tend to cause it to move; alternatively, if the body is already moving, then an applied force will tend to alter its motion (either in magnitude or in direction).

There are cases when a force is applied to a stationary object or structure and no perceptible change in motion occurs. In these cases the applied force is insufficient to overcome the inertia (resistance to motion) of the object; in other words, the frictional resistance to motion exerted on the object is greater than the applied force. If the object on which the applied force acts is fixed to a rigid structure, then again no motion is produced.

In these cases, forces equal and opposite to the applied forces resist the applied force and produce equilibrium. These internal resisting forces will act over the area which is resisting the applied force and the resisting force divided by the resisting area is known as the *stress*.

$$\text{Stress} = \frac{\text{resisting force}}{\text{resisting area}}$$

$$= \frac{\text{applied force or load}}{\text{resisting area}} \quad \text{N/m}^2$$

(Note that for convenience, small areas are often quoted in mm^2. A stress in N/mm^2 is numerically equal to the stress in MN/m^2, since both numerator and denominator are related by a factor of 10^6.)

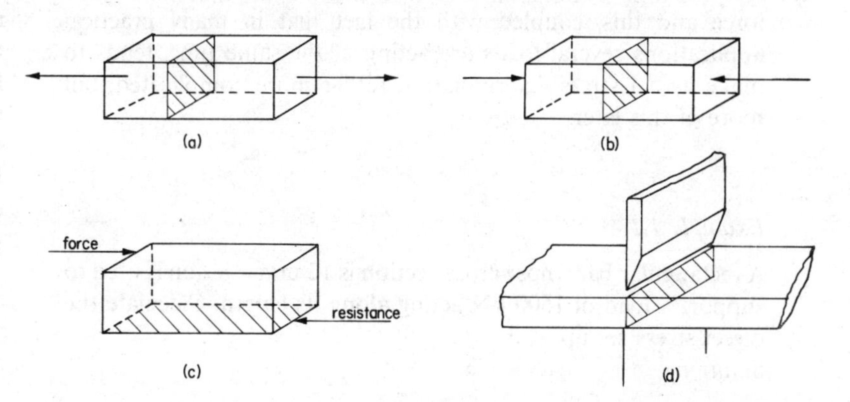

Figure 1.1 (a) A block subjected to a tensile force with its resisting area; (b) a block subjected to a compressive force with its resisting area; (c) a shear force applied to a block being resisted by an equal and parallel force acting on an area parallel to the plane of the applied force; (d) a guillotine blade 'shearing' metal

The area which resists the applied force will depend on the method of application of the applied force. When a tensile force is used (figure 1.1a) which pulls or tends to stretch an object, the area which resists the force is the cross-sectional area measured at right-angles to the line of application of the applied force. When a compressive force (figure 1.1b) tending to push or shorten the object (opposite to a tensile force) is used, the resisting area is the same, again measured at right-angles to the line of action of the force.

Tensile and compressive stresses are sometimes known as direct stresses and are represented by the Greek letter sigma (σ).

When a shearing force is used (figure 1.1c), tending to cause the body to slide or slip from its base, the area which resists the force is the area which lies parallel to the line of action of the force. Practical applications of shear are to be found in rivetted joints, punching and cropping of metals and twisting of shafts.

It should be noted that, since the area resisting the load is deemed to be unchanged by the load, the stress is proportional to the load.

In our later studies we shall see that a tensile load may produce a shear stress acting on a plane inclined to the line of action of the force and this, coupled with the fact that in many practical applications several loads are acting at the same time, tends to make actual stress determination rather more complicated, but more of this later.

Example 1.1

A rectangular bar whose cross-section is 12 mm × 8 mm is used to support a load of 1500 kN acting along its length. Calculate the direct stress set up.

Solution

$$\text{Stress, } \sigma = \frac{\text{load}}{\text{area}}$$

$$= \frac{1500 \times 10^3}{12 \times 8} = 15625 \text{ N/mm}^2 = 15625 \text{ MN/m}^2$$

$$= 15.625 \text{ kN/mm}^2$$

Example 1.2

A rivetted joint between two flat bars uses two rivets each 8 mm in diameter and carries a load of 1 MN. Calculate the shear stress set up in the rivets.

Solution

$$\text{Shear stress} = \frac{\text{load}}{\text{area}} = \frac{1 \times 10^6 \times 4}{2(\pi 8^2)}$$

$$= \frac{4 \times 10^6}{402}$$

$$= 9950 \text{ N/mm}^2 = 9.950 \text{ kN/mm}^2$$

1.2 STRAIN

In the cases under discussion, where the application of a force does not produce perceptible motion, in addition to setting up stresses the force will tend to deform or strain the object on which it acts. For small loads, which produce low stress values, the deformations will be very small and require accurate measuring techniques (for example, extensometers or strain gauges). These deformations will be elastic, that is, they will disappear when the load is removed. The deformations will be proportional to the applied load (or force) and proportional to the original dimensions (figure 1.2).

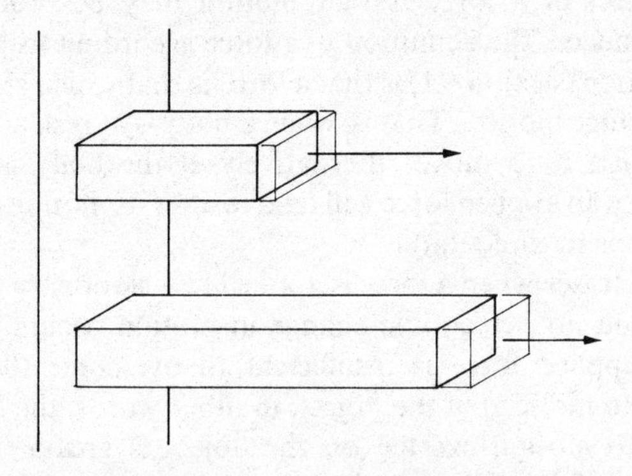

Figure 1.2 Loads of the same magnitude applied to two bars of the same material and of the same cross-sectional area will produce different deformations proportional to the original lengths of the bars

To express deformation in a form which is proportional to the applied load or force only, the term *strain* is used, which is defined as the actual deformation produced divided by the original dimension on which the force is acting. Thus

$$\text{strain} = \frac{\text{deformation produced}}{\text{original dimensions}} \; \frac{\text{mm}}{\text{mm}} \text{ or } \frac{\text{m}}{\text{m}}$$

therefore strain is a ratio and has no units.

The direction in which the deformation is produced is always in line with and is measured in the direction of the applied force, but the original dimension which is used in the strain formulae will depend on the condition of loading.

In the case of direct loading, that is, tensile or compressive loading, the original length or dimension is measured in line with the direction of loading and also in the direction of deformation; but in the case of shear loading the original length or dimension which is deformed is at right-angles to the line of action of the force and hence at 90° to the deformation (figures 1.3 and 1.15). The symbol for strain is the Greek letter epsilon (ε).

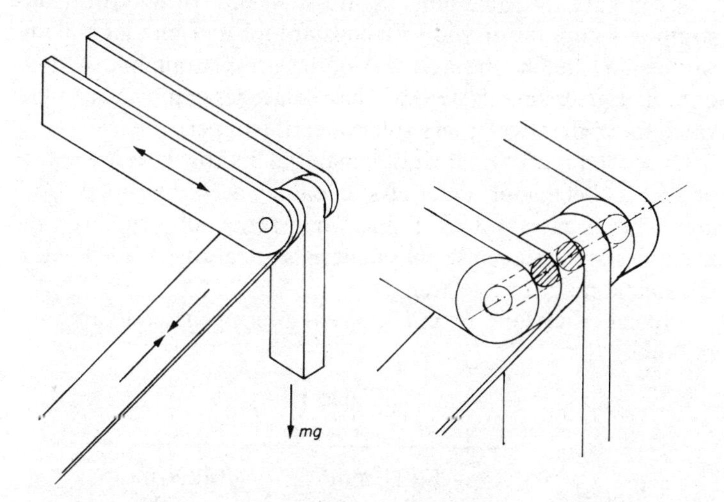

Figure 1.3 A load carried by a simple jib crane causes a tensile force along the length of the upper member (tie), imposes a compressive load along the length of the lower member (strut) and imposes a condition of shear in the pin. Note: the pin is said to be in a condition of double shear since the area resisting the shear force is twice the area of the pin

Example 1.3

A tie bar consisting of a rod of diameter 12 mm and length 1.5 m carries a tensile load of 800 kN, which produces an extension of 12 mm. Determine the stress and strain produced.

Solution

$$\text{Stress} = \frac{\text{load}}{\text{area}}$$

$$= \frac{800 \times 10^3 \times 4}{144\pi} = 7074 \text{ N/mm}^2 = 7074 \text{ MN/m}^2$$

$$\text{Strain} = \frac{x}{l} = \frac{12}{1.5 \times 10^3} = 0.008$$

1.3 HOOKE'S LAW AND THE MODULUS OF ELASTICITY

We have seen that, for smaller loads, and during the earlier stages of the application of gradually applied heavy loads, the material behaves elastically and both stress and strain are proportional to the applied load or force. This means that the graphs of stress against load and of strain against load will be straight lines indicating their linear relationship (figure 1.4).

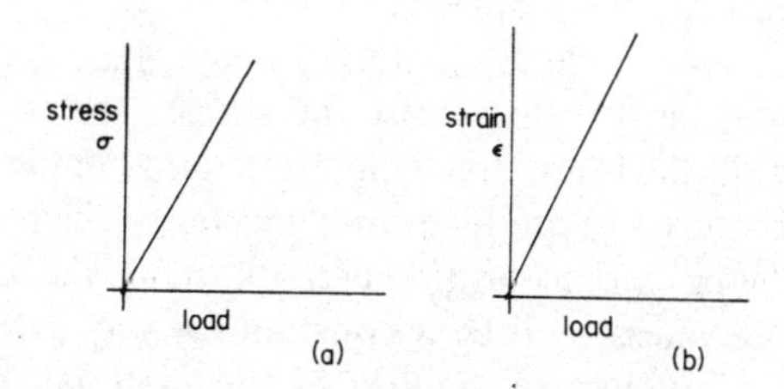

Figure 1.4 The linear relationship between (a) stress and load, where stress = load/area; (b) strain and load, where strain = x/l

Under these circumstances the material is said to follow *Hooke's law* (after Robert Hooke[1]), which states that the extension produced in an elastic material is directly proportional to the load which produced it.

This is normally represented graphically by combining the

above mentioned diagrams to give the load–extension graph shown in figure 1.5. This can easily be converted into a stress–strain graph by dividing the vertical load scale by the cross-sectional area, which will be constant, and by dividing the horizontal extension scale by the original length, which will also be constant; the shape of the graph remains unaltered.

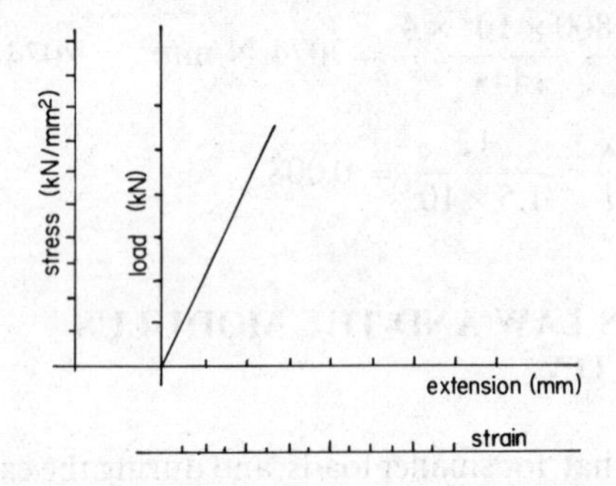

Figure 1.5 The load–extension and stress–strain relationships may be drawn on the same graph using different-scaled axes

It should perhaps be noted that this graphical representation does not follow strict mathematical practice, since the independent variable—the load—is plotted vertically.

The behaviour of different materials under these elastic loading conditions will be determined by the stiffness or rigidity of the material, that is, the more stiff or rigid a material is, the greater will be the load required to produce unit deflection. Different materials will have different slopes or gradients on the stress–strain graph. The slope (steepness) will be a constant for any given material, indicating the stiffness or rigidity of the material, and will be represented by the tangent of the angle of inclination to the horizontal.

When a material obeys Hooke's law, the ratio of load to deformation is constant; but

$$\text{load} = \text{stress} \times \text{area}$$

and

$$\text{deformation} = \text{strain} \times \text{length}$$

Since

$$\text{load/deformation} = \text{constant}$$

$$\frac{\text{stress} \times \text{area}}{\text{strain} \times \text{length}} = \text{constant}$$

but since area and length are constant for a given testpiece

$$\frac{\text{stress}}{\text{strain}} = \text{a constant denoted by } E$$

where E is the modulus of elasticity or Young's modulus of elasticity (after Thomas Young[2]).

E will have the same units as stress, N/mm^2 or MN/m^2 (since strain is a pure ratio), will be a constant for a given material and will be indicated by the steepness of the stress–strain line. A stiffer or more rigid material (steel) will have a steeper graph and a higher value for E than will a less stiff material (copper).

Once the value of E for a given material is known, we are able to predict its behaviour under elastic loading according to Hooke's law. This is most valuable since in practice all structures and machines are designed so that materials are always loaded within the elastic range (see section 1.9).

Typical values for E for common engineering materials are given in table 1.1.

Table 1.1

Material	E(kN/mm^2 or GN/m^2)	G(kN/mm^2 or GN/m^2)
Steel	200	80
Cast iron	100	42
Aluminium	70	26
Brass	90	35
Bronze	82	

Example 1.4

Each leg of a structure carries a mass of 800 kg whose weight acts axially along its length, which is 0.6 m. If the cross-sectional area of each leg is 180 mm^2 and the modulus of elasticity for the material is 200 GN/m^2, calculate the stress set up in each leg and the amount by which the leg shortens under load.

Solution

$$\text{Load} = 800 \times 9.81 = 7848 \text{ N}$$

$$\text{Stress} = \frac{\text{load}}{\text{area}} = \frac{7848}{180} = 43.6 \text{ N/mm}^2$$

Since E = stress/strain

$$\text{strain} = \frac{\text{stress}}{E} = \frac{43.6}{200 \times 10^3} = 0.000218$$

$$\text{strain} = \frac{x}{l}$$

therefore

$$\begin{aligned} x &= \text{strain} \times l \\ &= 0.000218 \times 0.6 \times 10^3 \\ &= 0.131 \text{ mm} \end{aligned}$$

1.4 ELASTICITY AND PLASTICITY

We have seen that for smaller loads during the earlier stages of loading the deformations produced are small, are proportional to the load and are not permanent, since the material is behaving elastically. The designed working stresses set up will be within this *elastic range* for the material.

As the load is increased it is found that the deformations produced become larger and are not proportional to the load. They also become permanent, the material remaining deformed even when the load is removed. On the load–extension graph (or stress–strain graph) the behaviour of the metal during this *plastic range* is represented by a curve (figure 1.6).

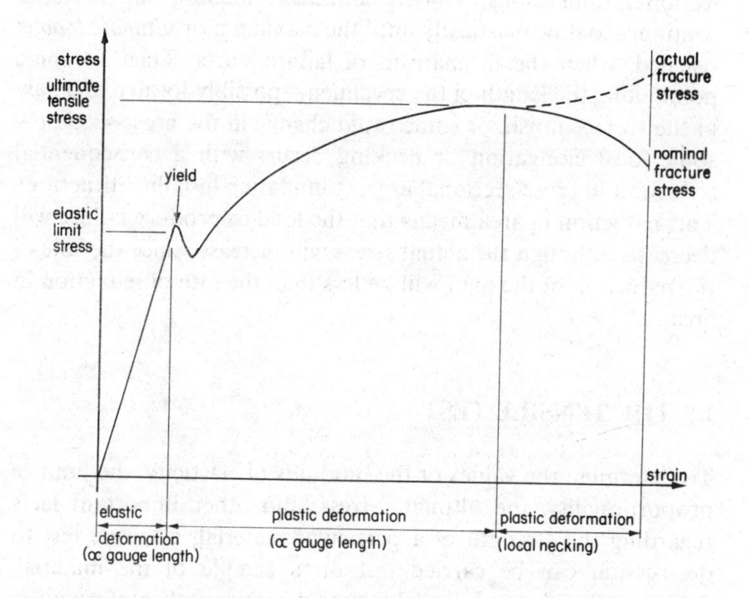

Figure 1.6 A stress–strain curve with a defined yield point indicating the main phases of extension and the main stress points

The end of the straight-line portion of the graph is indicated by the *limit of proportionality* and the material no longer obeys Hooke's law but remains elastic. The *elastic limit*, which may coincide with the limit of proportionality or be just above it, indicates the point in the loading where permanent deformations are produced.

In a few materials (such as mild steel) a quite clear distinction is shown when the material ceases to behave elastically and the material enters the plastic range, due to yielding (an increase in deformation with little or no increase, or even a reduction, in loading) brought about by the slipping of crystal planes. On more ductile materials this distinction may not be so clearly indicated.

During the first part of the plastic loading, up to the ultimate load or stress point, the increasing rate of deformation occurs over the whole length of the specimen under load. This increased deformation is deemed to have no appreciable effect on the cross-sectional dimensions. Under continued loading the material continues to flow plastically until the maximum or *ultimate load* is reached, when the mechanism of failure starts. Then, at some point along the length of the specimen—possibly located at a flaw in the surface finish, or some rapid change in the cross-section—some local elongation or necking occurs with a consequential reduction in cross-sectional area, culminating in failure (fracture). This reduction in area means that the load to produce failure will decrease, although the actual stress will increase, since the rate of the reduction of the load will be less than the rate of reduction in area.

1.5 THE TENSILE TEST

To determine the values of the modulus of elasticity, the limit of proportionality, the ultimate stress, and other important facts regarding the strength of a particular material, a tensile test to destruction can be carried out on a sample of the material. Additionally this test will enable us to determine other information regarding the ductility of the material, that is, the ability of the material to be drawn or stretched into a wire.

The normal commercial tensile test is carried out on a large hydraulically operated machine capable of exerting forces up to 250 kN. In the laboratory a similar test can be carried out on small testpieces on a small manually operated machine, known as a Hounsfield tensometer, to determine limited information about the properties of the material.

The normal specimens used are prepared by turning a known length to a specified diameter. To ensure that results may be usefully compared no matter where the test is carried out, the lengths of the specimens and their diameters are laid down by the British Standards Institution in BS 18. This Standard recommends that the gauge length shall be related to the diameter and also specifies the sizes for strip specimens used to test plate material.

BS 18: Methods of Tensile Testing of Metals

Part 1: 1970 Non-ferrous metals; Part 2: 1971 Steel (general); Part 3: 1971 Steel sheet and strip; Part 4: 1971 Steel tubes

Round Testpieces

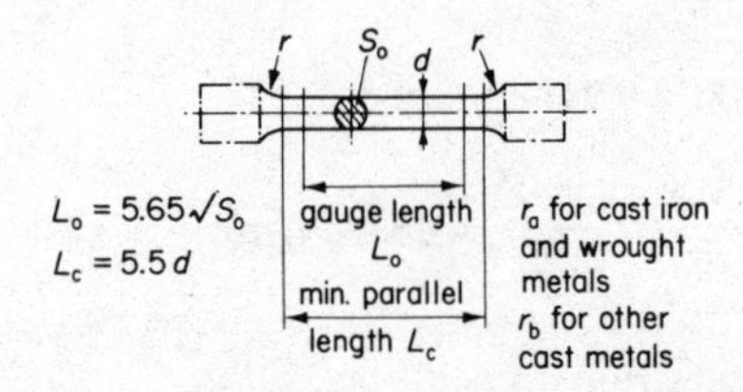

Dimensions of Circular-section Testpieces

S_0 (mm^2)	d (mm)	L_o (mm)	L_c (mm)	r_a (mm)	r_b (mm)	Tolerance on d ($\pm$mm)
200	15.96	80	88	15	30	0.08
150	13.82	69	76	13	26	0.07
100	11.28	56	62	10	20	0.06
50	7.98	40	44	8	16	0.04
25	5.64	28	31	5	10	0.03
12.5	3.99	20	22	4	8	0.02

Note Total extension x is made up of two component extensions: a = general extension proportional to gauge length l, that is, $b \times l$; b = local extension proportional to $\sqrt{}$cross-sectional area, that is, $c\sqrt{A}$. Then

$$x = bl + c\sqrt{A}$$

$$\%\text{ elongation} = \frac{bl + c\sqrt{A}}{l} \times 100$$

$$= 100(b + c\sqrt{A})$$

Part 1 Clause 2.3.1: Non-proportional testpieces used for plates and strips shall conform to tabulated dimensions.
Part 1 Clause 2.3.2: Machined testpieces for cast irons, etc., shall conform to tabulated dimensions. For wrought testpieces of uniform cross-section, gauge length = $5.65\sqrt{S_0}$.
Part 1 Clause 2.4: Tubes may be tested by plugging the ends or by strips cut longitudinally as tabulated.
Part 2 Clause 5.3: Testpieces that are geometrically similar and have a specified relationship between gauge length and cross-sectional area are known as proportional testpieces. For circular cross-sections the internationally accepted relationship is $L_0 = 5.65\sqrt{S_0}$, giving $L_0 = 5d$.
Part 2 Clause 6.3: Separation of the testpiece from the sample shall cause minimum heating and deformation.
Part 2 Appendix A: Testpieces shall not be cold straightened if a yield stress is to be determined. If the material specification permits cold straightening, a proof stress > 5% of extension shall be specified.

The specimen may be fitted with an extensometer, which is a device for measuring accurately the small elastic deformations which are produced during the early stages of the test and must be recorded to determine the modulus of elasticity. Various types of extensometer are available but the basic principle of operation is the same. The devices are attached to two points on the specimen separated by a distance given by the gauge length of the specimen. The relative movement of these two points during the elastic loading is multiplied and calibrated so that the extension can be recorded.

The specimen and extensometer are loaded into the tensile testing machine and gripped by means of split tapered jaws, threaded shackles or by split-chucks; care must be taken to ensure accurate alignment to produce pure tensile loading. The straining mechanism is operated to apply the load to the specimen and the load-measuring device indicates the load being applied at any particular moment. The loads and corresponding extensions are recorded. During the early stages of loading these are proportional to each other, giving a straight-line graph. At the end of this linear loading phase the extensometer is removed and the larger plastic extensions are recorded by a vernier device on the tensile testing machine. The results from this straight line enable the value of the modulus of elasticity to be determined and at the end of this phase the limit of proportionality and elastic limit can be fixed.

With mild steel the end of the elastic range is clearly marked by the *yielding*, caused by crystal dislocation within the material. In ductile material, such as copper, with a large plastic deformation, the beginning of the plastic range is not so clearly indicated. In these cases it is normal to use a *percentage proof stress* figure instead of the elastic limit stress (or limit of proportionality stress). This is the stress which produces a permanent deformation equal to a specified percentage of the original gauge length. It is determined by marking a point on the horizontal extension axis to represent a certain percentage extension, for example 0.1 per cent strain (figure 1.7). Thus

$$0.1\%\ \text{strain} = \frac{\text{extension}}{\text{length}} = \frac{1/1000\ (\text{gauge length})}{(\text{gauge length})}$$

therefore

$$\text{extension} = \frac{1}{1000}\,(\text{gauge length})$$

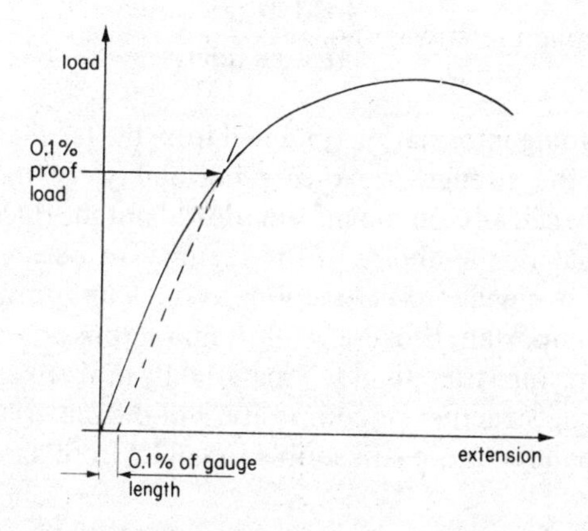

Figure 1.7 A load–extension graph for a ductile material with no clearly defined yield point, showing the determination of the 0.1% proof load

A line passing through this point is drawn parallel to the initial straight-line portion of the graph to intersect the curve at a point which gives the 0.1 per cent proof load.

The earlier part of the plastic loading occurs over the whole length of the specimen and so is proportional to the gauge length but does not affect the diameter.

The maximum load applied divided by the original area will give the ultimate tensile stress of the material. After this point further plastic elongation occurs in a local region giving a reduction in the cross-sectional area resisting the load. This causes a reduction in the load which reduces until fracture occurs. The nominal fracture stress is obtained by dividing the load at fracture by the original area, thus

$$\text{nominal fracture stress} = \frac{\text{fracture load}}{\text{original area}}$$

However, during this phase of failure, although the load is decreasing, the greater reduction in cross-sectional area means that the actual stress will continue to increase and the true stress at fracture is given by

$$\text{actual fracture stress} = \frac{\text{load at fracture}}{\text{area at fracture}}$$

The foregoing information obtained from the tensile test relates directly to the strength (load-carrying ability) of the material. However, we can also obtain information about the ductility of the material (that is, the ability of the material to be drawn out in tension to a smaller cross-section, as in wire drawing). This property is important because, while it may be possible to improve the load-carrying strength of the material by heat treatment or by modifying the structure or composition of the material, this may make the material more susceptible to failure from shock loading or vibration.

Ductility of materials is usually measured by the percentage elongation figure and the percentage reduction in area figure, both of which are obtained after a tensile test to destruction.

The actual elongation produced after fracture is determined by placing the two halves of the specimen together and measuring the increase in length and a percentage figure is determined.

$$\text{Percentage elongation} = \frac{\text{elongation}}{\text{original length}} \times 100$$

$$= \frac{\text{final length} - \text{original length}}{\text{original length}} \times 100$$

Care must be taken when using this figure because, as we have seen, the total extension is made up of two parts

(1) the small elastic deformation and the early plastic deformation, both of which occur over the whole gauge length and will be proportional to the gauge length;

(2) the local extension due to the 'necking' process, which is independent of the gauge length (being dependent on the square root of the cross-sectional area).

It is therefore important that whenever the percentage elongation figure is used, the gauge length on which it is based be quoted.

The percentage reduction in area figure is given by

$$\frac{\text{reduction in area}}{\text{original area}} \times 100$$

that is

$$\frac{\text{original area} - \text{final area}}{\text{original area}} \times 100$$

The final area is obtained by calculation after measuring the diameter across the site of fracture.

Examination of fractures which are produced as a result of the tensile test can be used to determine the mode of failure, the condition of the material (by the grain structure) and the presence of imperfections and impurities (by blowholes or slag inclusions).

There are many different fractures that can occur, depending on the material and its condition. However, they may be grouped broadly as (1) cup and cone fractures or (2) shear plane fractures (see figure 1.8).

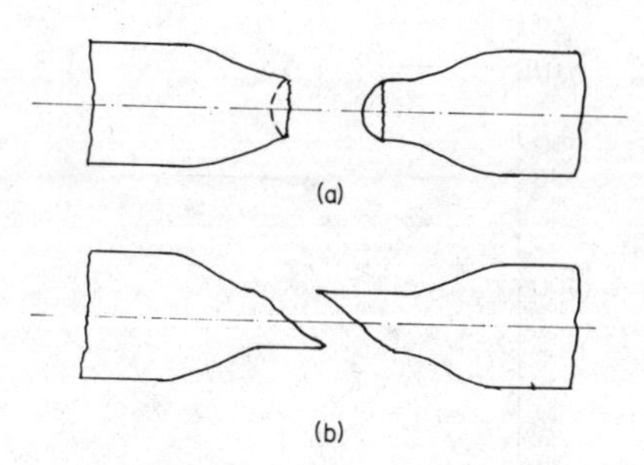

Figure 1.8 Typical fractures after a tensile test: (a) cup and cone fracture resulting from failure in tension; (b) shear plane fracture resulting from a tensile test specimen failing in shear

The cup and cone fracture will occur in ductile materials and is a common characteristic of mild steel. Various forms of this type of fracture may occur according to the alloying elements present.

Materials that are weaker in shear than in tension may fail due to shear stress which will be set up on any plane inclined at any angle to the tensile pull. This shear stress is usually a maximum at an angle of 45°. Wrought aluminium and duralumin are typical examples of materials prone to this type of fracture.

Example 1.5

A light alloy testpiece 10 mm in diameter and gauge length 50 mm gave the following results when tested to destruction.

Load at yield point 25 kN
Maximum load 40 kN
Extension at yield point 0.25 mm
Diameter at fracture 8.5 mm

Determine (a) stress at yield point, (b) Young's modulus of elasticity for the material, (c) ultimate tensile stress and (d) percentage reduction in area.

Solution

$$\text{Initial cross-sectional area} = \frac{\pi}{4}d^2$$

$$= 78.54 \text{ mm}^2$$

$$\text{Stress at yield point} = \frac{\text{load at yield point}}{\text{area}}$$

$$= \frac{25 \times 10^3}{78.54} = 318.3 \text{ N/mm}^2$$

$$= 318.3 \text{ MN/m}^2$$

$$\text{Strain at yield point} = \frac{x}{l}$$

$$= \frac{0.25}{50} = 0.005$$

$$\text{Modulus of elasticity} = \frac{\text{stress}}{\text{strain}} \text{ (within the elastic limit)}$$

$$= \frac{318.3}{0.005} = 63\,660 \text{ N/mm}^2$$

$$= 63.66 \text{ kN/mm}^2 \text{ or } 63.66 \text{ GN/m}^2$$

$$\text{Ultimate tensile stress} = \frac{\text{ultimate or max. load}}{\text{original area}}$$

$$= \frac{40 \times 10^3}{78.54}$$

$$= 509.3 \text{ N/mm}^2 \text{ or MN/m}^2$$

$$\text{Final area} = \frac{\pi}{4}d^2 = \frac{\pi}{4}(8.5)^2$$

$$= 56.74 \text{ mm}^2$$

$$\%\text{ reduction in area} = \frac{\text{(original} - \text{final) area}}{\text{original area}} \times 100$$

$$= \frac{78.54 - 56.74}{78.54} \times 100$$

$$= \frac{21.8}{78.54} \times 100$$

$$= 27.75\,\%$$

Example 1.6

During a test to destruction carried out on a mild steel bar, original diameter 24 mm, gauge length 250 mm, the following results were obtained.

Load (kN)	*Extension* (mm)
11.95	0.030
19.9	0.056
28.8	0.081
40.25	0.118
49.8	0.140
61.7	0.173
70.7	0.198
79.7	0.203
91.8	0.254
100	0.274
110.6	0.305
120	0.335
129.5	0.3658
139.5	0.68 (yield point)
198.8	max. load

After the test the diameter at fracture was found to be 15 mm and the length was 320 mm.

Draw the load–extension graph and determine (a) elastic limit stress, (b) maximum stress, (c) modulus of elasticity, (d) percentage extension in length, (e) percentage reduction in area, (f) 0.1 per cent proof stress.

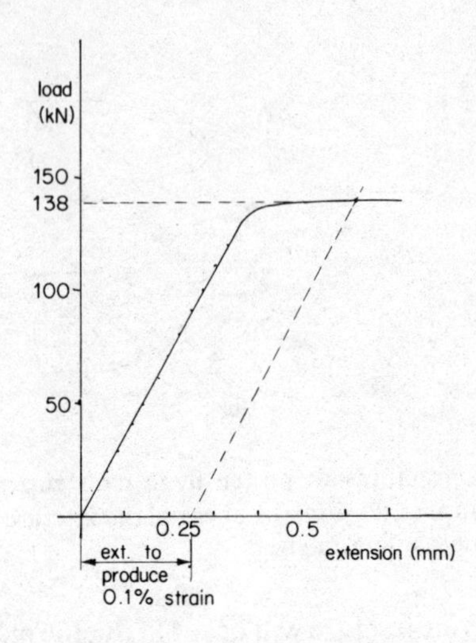

Figure 1.9

Solution The graph is as shown in figure 1.9.

(a) Elastic limit $= \dfrac{139.5 \times 10^3}{(\pi/4)\,(24)^2} = \dfrac{139.5 \times 10^3}{452.4} = 309\ \text{N/mm}^2$

(b) Maximum stress $= \dfrac{198.8 \times 10^3}{452.4} = 439\ \text{N/mm}^2$

(c) Modulus of elasticity $= \dfrac{\text{stress}}{\text{strain}}$ (within elastic limit)

$$= \frac{\text{load}}{\text{area}} \times \frac{\text{length}}{\text{extension}}$$

$$= \frac{100 \times 250}{452.4 \times 0.274}$$

$$= 201.2\ \text{kN/mm}^2 \text{ or GN/m}^2$$

(d) Percentage extension $= \dfrac{\text{extension}}{\text{original length}} \times 100$

$$= \frac{320 - 250}{250} \times 100$$

$$= \frac{70}{250} \times 100 = 28\%$$

(e) Percentage reduction in area

$$= \frac{\text{reduction in area}}{\text{original area}} \times 100$$

$$= \frac{\text{original area} - \text{final area}}{\text{original area}} \times 100$$

$$= \left[\frac{452.4 - (\pi/4)(15)^2}{452.4}\right]100$$

$$= \left(\frac{452.4 - 176.7}{452.4}\right)100$$

$$= \frac{275.7}{452.4} \times 100 = 60.9\%$$

(f) Extension to produce 0.1 % strain

$$0.001 = \frac{x}{l}$$

therefore

$$x = 0.001 \times 250 = 0.250$$

Load to produce 0.250 mm permanent extension (see figure 1.9) = 138 kN, thus

$$0.1\% \text{ proof stress} = \frac{\text{load}}{\text{area}} = \frac{138 \times 10^3}{452.3}$$

$$= 305 \text{ N/mm}^2 = 305 \text{ MN/m}^2$$

1.6 PROPERTIES OF MATERIALS AND STRESS–STRAIN CURVES

When a designer is choosing a suitable material for a specific function, he has to consider many different requirements: the suitability of the material for the likely forming methods to be employed, the reaction of the material to the type and method of application of the load, the reaction of the material to the environmental situation, and economic factors.

The mechanical properties of the material that will have to be taken into consideration include strength, toughness, ductility and hardness. We have already met some of these properties—strength, the load-carrying ability of the material; ductility, the ability of the material to be drawn into wire—and we have seen how the stress–strain curve can be used to illustrate these properties. These properties may be modified by heat treatment or alloying when necessary.

Most materials in their pure state (for example, aluminium and copper) are weak (that is, have poor resistance to load) with low elasticity and considerable plasticity (figure 1.10). Materials which are called brittle (that is, have poor resistance to shock loading) exhibit low plasticity with some elasticity (figure 1.11). The rigidity and stiffness of materials is indicated by the steepness of the curve (and therefore by higher values of E) (figure 1.12). More ductile materials exhibit considerable plasticity (figure 1.13). The effect of alloying on the shape of stress–strain curves is illustrated in figure 1.14, which shows the effect of carbon on the strength and behaviour of steel.

Malleability is another property which warrants consideration; this is the ability of the material to be beaten into a sheet.

A further important property of materials which has to be considered by the designer when considering the possible use of a material for a particular function is the hardness of the material. Hardness is defined as the resistance of the material to scratching or to indentation. Again this property of the material can be modified by heat treatment, in the case of ferrous materials, and by work-hardening and exposure to the atmosphere, in the case of non-ferrous (copper and aluminium) alloys.

There are many methods of testing and measuring the hardness

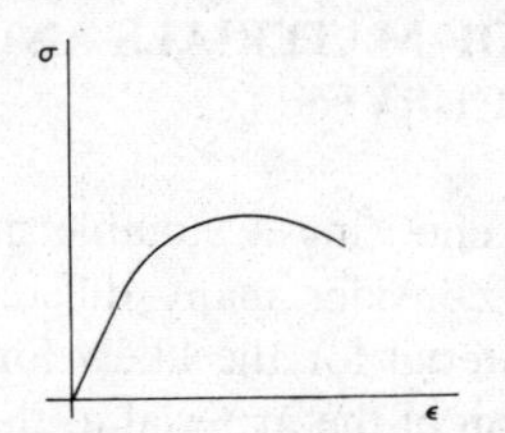

Figure 1.10 Stress–strain curve for a material with low elasticity and some plasticity—pure material

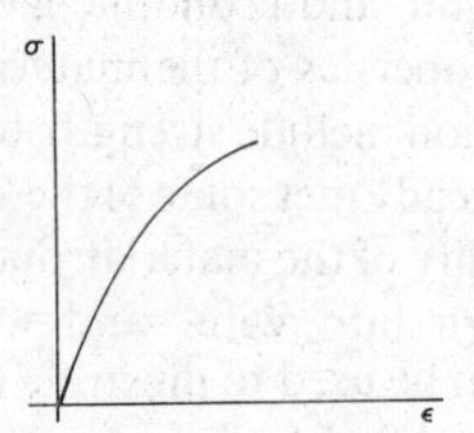

Figure 1.11 Stress–strain curve for a material with some elasticity but little or no plasticity—brittle material

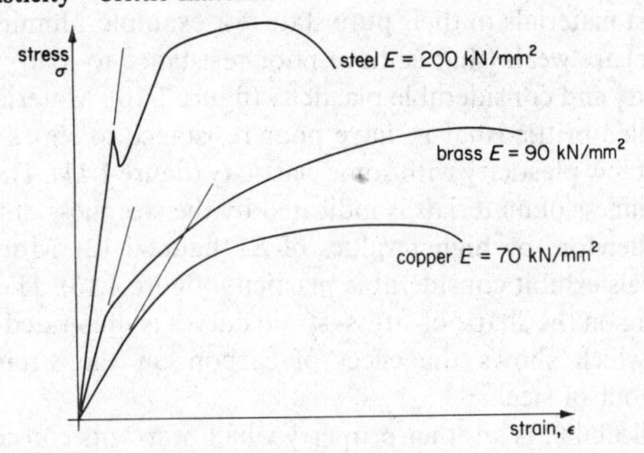

Figure 1.12 Different materials showing different gradients for the elastic range, depending on conditions

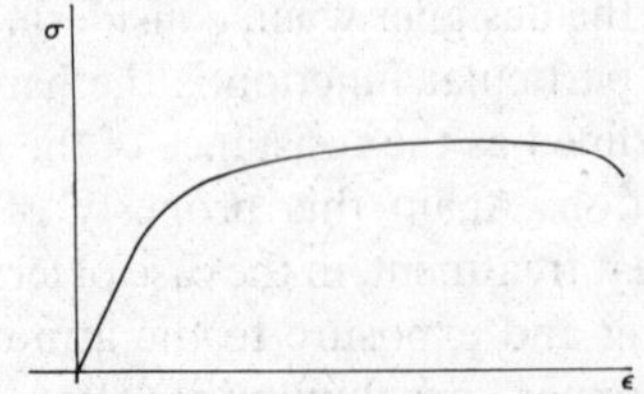

Figure 1.13 Ductile materials showing considerable plasticity, as indicated by this stress–strain curve (see also figure 1.12)

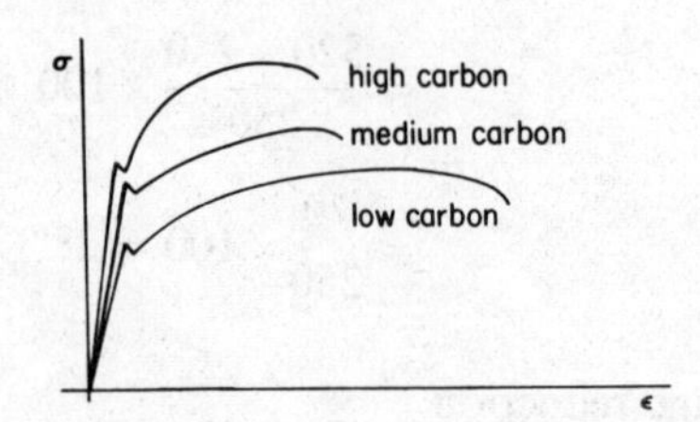

Figure 1.14 Increasing amounts of carbon in steel, besides making the steel more amenable to heat treatment, make the steel stronger but reduce the ductility, as indicated by these stress–strain curves

of materials, the most common of which are the Vickers, the Brinell and the Rockwell methods. In these tests, a small hardened steel ball or a diamond pyramid is pressed into a prepared surface of the material to be tested, using a known load, and then the diameter or diagonal of the indentation is measured.

1.7 INTRODUCTION TO PLASTICS MATERIALS

So far we have considered the stress–strain curves for metallic materials (both ferrous and non-ferrous) but the application of various plastics materials in engineering is on the increase and stress–strain curves can be used to illustrate their reactions under load. Polymeric materials (plastics) can be classified into two categories: (1) thermoplastic and (2) thermosetting: the distinction lies in their reaction to heat. Thermoplastic materials soften on the application of heat, returning to their rigid state on cooling; this cycle of 'softening and hardening' can be repeated indefinitely. Materials which behave in this manner include nylon, polyethylene and polystyrene. Thermosetting materials, which include phenolics, epoxies and polyesters, soften once on heating but when rigid they cannot be remelted.

The behaviour of plastics under load is affected by their temperature, and although in general their strength is not high as compared to metals (ultimate tensile strength 35 to 70 MN/m² compared to 400 to 600 MN/m² for mild steel), their many other advantages are likely to ensure that the world consumption of plastics will increase at a faster rate than the consumption of metals. Although the economic advantages of plastics have been eroded by the increasing price of oil, the general rate of inflation will ensure that plastics will be able to compete economically with metals in the future.

The other advantages of plastics include lightness, toughness, good resistance to corrosion and to fatigue, low coefficients of friction, ease of forming, and pleasing appearance. The strength of plastics materials can be improved by the use of fillers or glass fibres and their hardness varies according to type—acrylic plastics being the hardest.

1.8 SHEAR STRESS AND SHEAR STRAIN—MODULUS OF RIGIDITY

We have considered the relationship between stress and strain under the application of tensile and compressive loads and have seen that, while the material is behaving elastically, stress and strain are proportional to each other, the ratio between them being a constant—the modulus of elasticity—for any given material.

We have seen that in the case of shear loading, where the applied load is resisted by an equal parallel force, the resisting area is in the same plane as the applied load. In addition the shear load will produce a deflection which may be measured in the same plane as that in which the load is applied. Again while the material behaves elastically (that is, the deflection totally disappears when the load is removed) the deflection will be proportional to the applied load and the original dimensions. In this case, however, because the resisting area is in the same plane as the load, and the original length is measured at 90° to the resisting area, the original length is measured at 90° to the plane in which the deformation occurs (figure 1.15). This shear strain, denoted by ε, is given by

$$\varepsilon = \frac{\text{deformation}}{\text{original length}}$$

(measured at 90° to the deformation).

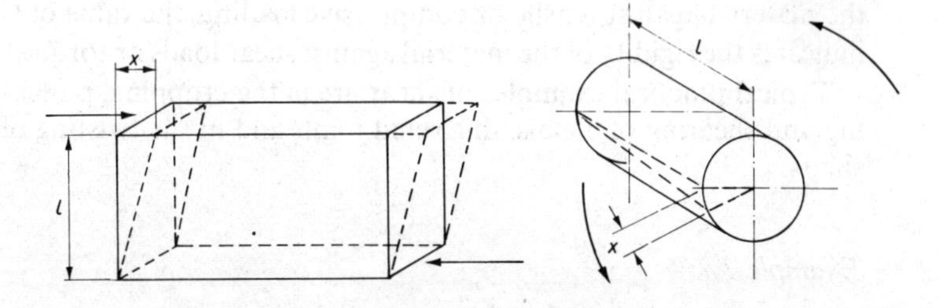

Figure 1.15 Under conditions of shear, the original length is measured at 90° to the direction of deformation produced

These small deflections appearing as the result of shear loads can be seen to disappear when the load is removed. Consequently the material is behaving elastically and Hooke's law again applies, that is, the ratio load/deformation is a constant. Now

$$\text{load} = \text{shear stress} \times \text{area}$$
$$\text{deformation} = \text{shear strain} \times \text{length}$$

therefore

$$\frac{\text{shear stress} \times \text{area}}{\text{shear strain} \times \text{length}} = \text{constant}$$

but since area and length are constant for a given specimen

$$\frac{\text{shear stress}}{\text{shear strain}} = G$$

where G is the modulus of rigidity for a given material (typical values for which are given in table 1.1).

Increased loading will result in larger deflections and it becomes apparent that these are not proportional to the applied load (or torque) and that they are part plastic, that is, part of the deformation is permanent. The material is now behaving plastically and failure will result.

Just as the value of E for a given material indicates the rigidity of the material against tensile or compressive loading, the value of G indicates the rigidity of the material against shear loads or torques.

Typical practical examples of shear are in the cropping, punching and shearing of metals, in riveted joints and in the twisting of shafts.

Example 1.7

Find the force required to shear a bar of metal 30 mm × 8 mm thick if the ultimate shear stress of the metal is 385×10^6 N/m^2.

Solution

$$\text{Area} = 30 \times 8 = 240 \text{ mm}^2$$
$$\text{Force} = \text{stress} \times \text{area} = 385 \times 240 = 92400 \text{ N}$$

Example 1.8

If a rivet has to carry a shear load of 100 kN and the maximum stress must not exceed 250×10^6 N/m^2, find the minimum diameter of the rivet.

Solution

$$\text{Area} = \frac{\text{load}}{\text{stress}} = \frac{100 \times 10^3}{250} = 400 \text{ mm}$$

therefore

$$d = \sqrt{\left(\frac{400 \times 4}{\pi}\right)}$$
$$= 22.5 \text{ mm}$$

Example 1.9

A torsion bar is subjected to a shear stress of 45 GN/mm^2. If the modulus of rigidity of the material is 80 GN/m^2, determine the strain set up.

Solution

$$\frac{\text{shear stress}}{\text{shear strain}} = G$$

$$\text{shear strain} = \frac{\text{shear stress}}{G}$$

$$= \frac{45}{80} = 0.5625 \text{ (no units)}$$

1.9 FACTOR OF SAFETY AND WORKING STRESSES

It is important that the designed working stresses always lie within the elastic limit for the material, so that, when the load is removed, any deformations or deflections so caused will also disappear. (It would be disastrous if the first time the maximum load were applied to a bridge the bridge took on a permanent deflection.)

When a designer chooses to use a certain material which is best suited for his purpose, he is able to refer to the material specification to find the value of the ultimate stress (tensile, compressive, or shear, depending on the loading conditions); in some cases he may refer to the yield or elastic limit stress. The working stress is related to this known value by a factor of safety, which will be based on many considerations including the designer's past experience and previous practice, where

$$\text{working stress} = \frac{\text{ultimate stress}}{\text{factor of safety}}$$

Among the factors that will influence the choice of the factor of safety and hence the size of the working stress are

(1) the relationship of the elastic limit and yield stress to the ultimate stress;

(2) the method of application of the load, for example,

gradually applied, suddenly applied or applied with impact (a suddenly applied load will produce an instantaneous stress figure of twice that nominally calculated for a gradually applied load);

(3) whether the load is constantly applied or whether it is a recurring load; this condition may cause the material to fail due to fatigue;

(4) if the load or stress is reversed, as may occur in a vibrating beam or shaft; this will also induce a fatigue failure;

(5) the consequences of failure: where human life may be put at risk as a result of the possible failure, a higher factor of safety will be used;

(6) environmental conditions: corrosion may produce stress concentrations which will lead to premature failure;

(7) economic considerations;

(8) length of working life;

(9) factor of ignorance.

Example 1.10

A steel bolt has an effective diameter of 15 mm. If the ultimate tensile stress for steel is 600 MN/m^2 and a factor of safety of 6 is used, calculate the maximum safe load that may be carried by the bolt.

Solution

$$\text{Working stress} = \frac{\text{ultimate tensile stress}}{\text{factor of safety}}$$

$$= \frac{600}{6} = 100\ \text{MN/m}^2$$

$$\text{area of bolt} = \frac{\pi d^2}{4} = \frac{\pi 15^2}{4}$$

$$= 176.7\ \text{mm}^2 = 176.7 \times 10^{-6}\ \text{m}^2$$

$$\text{safe working load} = \text{stress} \times \text{area}$$

$$= 100 \times 10^6 \times 176.7 \times 10^{-6}$$

$$= 17670\ \text{N} = 17.67\ \text{kN}$$

1 Robert Hooke (1635–1703) was an English scientist who specialised in mechanics and is remembered for Hooke's law, which can be related to a spring or any other elastic body, and for Hooke's universal joint. Despite his specialisation he also did notable work in the fields of physics (light), chemistry (early combustion theory), physiology (respiration) and biology (cellular structure).

He was the son of the vicar of Freshwater in the Isle of Wight. He was educated at Westminster and Oxford. He was not wealthy and did not enjoy good health. He was compelled to obtain work to support himself while pursuing his scientific interests and he worked as an assistant to Sir Robert Boyle.

In 1662 he was appointed the first Curator of Experiments to the Royal Society and held this post until his death in 1703. It was during this time that he worked with Newcomen who produced the first workable engine; he also worked on celestial motion and gravitational attraction, and time-keeping devices and formulated his 'law'. Unfortunately from his correspondence and dealings with Newton (who was elected President of the Royal Society after the death of Hooke) it would seem that the two men did not enjoy the happiest of relationships.

Hooke was also a noted surveyor and architect and played a part in the rebuilding of the City of London after the Great Fire in 1666, along with Sir Christopher Wren.

2 Dr Thomas Young (1773–1829) was born at Milverton, Somerset, and was a child genius—he could read at the age of two and could speak many languages by the age of twelve. He studied medicine in England and Europe but had many interests outside this field. He was a notable Egyptologist and is famed in this field for his translation of hieroglyphics. He was a notable natural philosopher and was professor of physics at the Royal Institution from 1801 to 1803.

During this time he studied the nature of light closely and disputed Newton's particle theory, supporting early work by Huygens and Grimaldi, propounding that light travelled in wave forms. He worked in Paris with Fresnel and found support from Frenchmen in his dispute with Newton's theory. He also used his medical knowledge in developing a theory by which the human eye was able to distinguish colours by using different nerve fibres to distinguish red, green and violet.

TO THE STUDENT

At the end of this chapter you should be able to

(1) recognise tensile compressive and shear forces
(2) define stress and strain and solve simple problems involving stress and strain
(3) draw graphs of force against extension and stress against strain for elastic materials
(4) define Young's modulus of elasticity for a material, relating it to the stiffness of materials
(5) solve problems involving stress, strain and Young's modulus of elasticity
(6) describe the requirements of a tensile test to destruction, carried out to British Standards, illustrating the results by complete load–extension or stress–strain curves for brittle and ductile materials
(7) evaluate from the results of the above tests: the elastic limit, limit of proportionality, ultimate tensile stress, percentage elongation, percentage reduction in area and proof stress
(8) define ductility, brittleness and hardness
(9) complete exercises 1.1 to 1.17.

EXERCISES

1.1 A casting of mass 200 kg is supported by a wire 2 mm in diameter. Calculate the stress produced in the wire.

1.2 A tie bar 4 mm × 12 mm carries a pull of 12000 N. Determine the stress set up in the bar.

1.3 A tube 20 mm outside diameter and 16 mm bore is used to support a compressive load of 4 tonnes (1 tonne = 1000 kg). Calculate the compressive stress set up.

1.4 A rivet 10 mm diameter is used as part of a joint and as such carries a shear load of 5000 kg. Determine the shear stress set up in the rivet.

1.5 A girder whose cross-sectional area is 425 mm^2 carries a compressive load. Determine the maximum value of this load if the maximum stress must not exceed 50 kN/mm^2.

1.6 A column has a rectangular cross-section and is 2.5 m high. It carries a vertical load which produces a strain of 0.00052. Determine the amount by which the column shortens.

1.7 A metal rod of square cross-section, side length 25 mm, is subjected to a tensile force of 110 kN. The length of the rod before loading was 3 m and afterwards it was found to be 3.013 m. Assuming that the elastic limit has not been exceeded, find the value of Young's modulus for the material.

1.8 A bar 20 mm diameter carries a load of 30 kN. If the bar is 1.5 m long, and Young's modulus for the material is 200 kN/mm^2, calculate the extension produced.

1.9 A bar 25 mm diameter and 2.5 m long carries a load which produces an extension of 2.25 mm. If E for the material is 180 kN/mm^2, determine the load.

1.10 Determine the force necessary to punch a 20 mm diameter hole in a brass sheet 4 mm thick if the ultimate shear stress is 110 MN/m^2. Determine the compressive force set up in the punch.

1.11 A strip of metal 50 mm wide is guillotined into short lengths. If the metal is 8 mm thick and the ultimate shearing stress is 340 MN/m^2, determine the force set up at the blade.

1.12 A hacksaw blade of section 18 mm × 1 mm is tightened up in its frame so that the maximum tensile stress set up is 14 N/mm^2. Calculate the force applied if the pin holes in the blade are 6 mm in diameter. Find the shearing stress set up in the pins.

1.13 In a belt drive mechanism the belt used is 60 mm wide and 6 mm thick and is capable of carrying 100 N per 10 mm of its width. Find the maximum safe load and the stress in the belt at this load.

1.14 A load of 20 tonnes is to be carried by four ropes of equal diameter, made from steel whose ultimate tensile strength is 580 MN/m^2. Using a factor of safety of 6, find the minimum permissible diameter of the ropes. If the ropes are 3 m long find the extension produced. ($E = 200$ GN/m^2)

1.15 A strut of 20 mm diameter is 0.5 m long and must not shorten by more than 0.01 mm. Determine the maximum compressive load it may carry. ($E = 200$ GN/m^2)

1.16 During a tensile test on a specimen, the following results were recorded.

Load (kN)	*Extension* (mm)
0	0.006
0.5	0.014
1.0	0.022
1.5	0.030
2.0	0.038
2.5	0.046
3.0	0.054
3.5	0.062
4.0	0.070
4.5	0.079
5.0	0.090
5.5	0.111
5.75	0.321
8.0	max load

Specimen diameter 14.3 mm, gauge length 50 mm. Final diameter at fracture 9.07 mm, final length 67.3 mm. Determine (a) stress at the limit of proportionality, (b) ultimate tensile stress, (c) percentage elongation, (d) percentage reduction in area and (e) modulus of elasticity.

1.17 A light alloy testpiece gave the following results when tested to destruction. Load at yield point 26.25 kN; extension at yield point 0.25 mm; maximum load 43.6 kN; diameter at fracture 8.9 mm; original diameter 10 mm; gauge length 50 mm. Determine (a) stress at the yield point, (b) ultimate stress, (c) percentage reduction in area and (d) modulus of elasticity.

NUMERICAL SOLUTIONS

1.1 1250 N/mm^2
1.2 250 N/mm^2
1.3 346.8 N/mm^2
1.4 625.63 N/mm^2
1.5 212 MN/mm^2
1.6 1.3 mm
1.7 41 kN/mm^2
1.8 0.716 mm
1.9 318 kN
1.10 27.6 kN
1.11 136 kN
1.12 168 N; 5.96 N/mm^2
1.13 600 N; 1.667 N/mm^2
1.14 25.4 mm; 1.45 mm
1.15 1256 N
1.16 (a) 247 MN/m^2, (b) 294 MN/m^2, (c) 33%, (d) 59%, (e) 87 GN/m^2
1.17 (a) 322 MN/m^2, (b) 537.8 MN/m^2, (c) 24%, (d) 64.5 GN/m^2

2 Statics—Structures

The object of this chapter is that the student shall appreciate the effect of forces and loads on structures in equilibrium (frameworks and beams) and be able to calculate the effects of these forces and loads.

2.1 FORCE AS A VECTOR

Force is a vector quantity because, for its complete definition, we must specify both its magnitude and direction. This means that we are able to represent a force by a vector, that is, a straight line whose length (to some suitable scale) represents the magnitude or size of the force and whose direction indicates the direction in which the force is acting (figure 2.1).

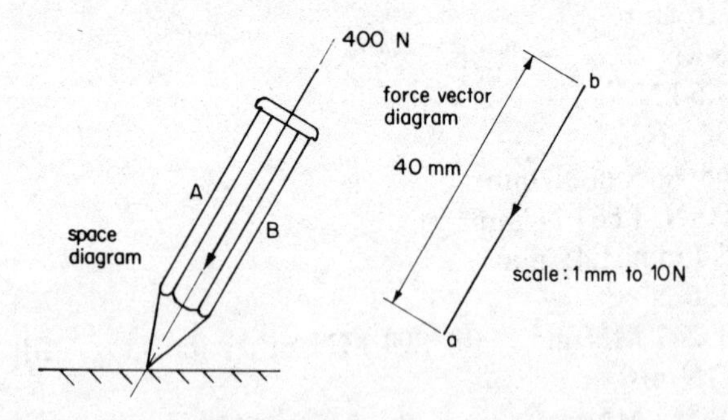

Figure 2.1 Vector representation of a force; note that the scale must always be stated on the vector diagram

The next point to appreciate is that, at any point or on any object, more than one force can be acting at the same time. This occurs in practice, for instance, at the tool point of a parting-off tool on a lathe, when there will be one force acting along the tool at 90° to the axis of the component and another force acting vertically downwards on the tool point (figure 2.2). On a roof structure or framework the gravitational attraction (weight) exerted on the roof covering will set up downward acting forces on the structure, while the effect of wind would be to impose a sideways acting force.

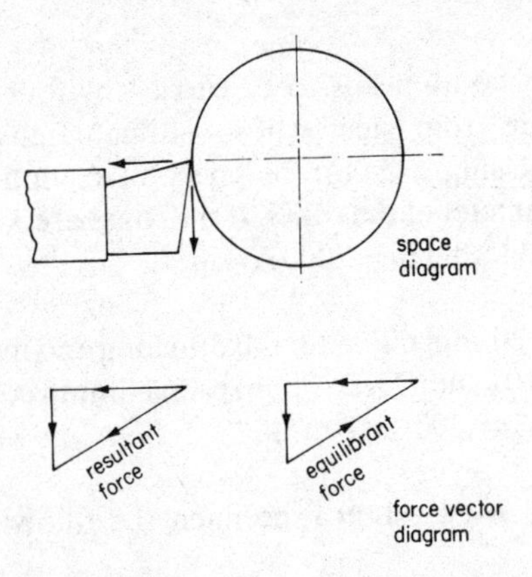

Figure 2.2 Two forces acting concurrently at the same point may be resolved to produce a resultant force giving the same effect; a force equal and opposite to this resultant will balance out the applied forces and produce equilibrium

If we combine the effect of any two forces, or any system of forces, we can replace the two (or more) component forces, or the system of forces, by a single *resultant force*, which will have the same total effect.

In reverse, it is possible to replace a single force by two (or more) *component forces*, which will produce the same total effect (figure 2.3).

When two or more forces acting at a point are opposed or balanced by a force which is equal in magnitude to, but opposite in direction to the resultant, then the point and system of forces is in *equilibrium*, and the force equal and opposite to the resultant is known as the *equilibrant force* (figure 2.2).

When an object is in equilibrium under a system of concurrent

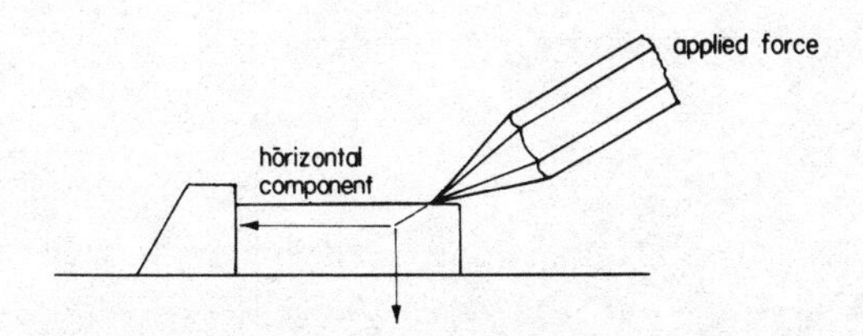

Figure 2.3 The single force applied by the chisel can be split into two component forces producing the same over-all effect; the horizontal force will tend to produce sliding along the bench while the vertical workpiece will increase the reaction between the component and the bench

and co-planar forces (that is, a system of forces acting in the same plane and all passing through the same point), then because there is no resultant force there is no tendency for the object to move (or to rotate).

We can probably best consider this idea of equilibrium, relating it to a force vector diagram as in figure 2.2, by considering the most common of all forces, that of gravity. Consider a mass, m, whose gravitational attraction or force, mg, is acting through its centre of gravity, being supported by two inclined wires in the same plane, as shown in figure 2.4a. This will give us a co-planar concurrent force system. The downward acting gravitational force mg is balanced by the vertical effect of the tensions in the wires. The tension in each wire may be resolved into horizontal and vertical components and the sum of the vertical components will be equal to downward vertical acting mg, as shown in figure 2.4b. It will be seen that by rearranging the vector triangles to form a parallelogram, as in figure 2.4c, a more convenient form of vector representation can be made, whereby the diagonal of the parallelogram is the sum of the vertical components, that is, the equilibrant of the gravitational force mg.

This is expressed by the theorem of the parallelogram of forces which states that if two forces acting at a point are represented in magnitude and direction by the two adjacent sides of a parallelogram, their resultant will be represented both in magnitude and direction by the diagonal of the parallelogram drawn from the point. Note that this parallelogram may be superimposed on the space diagram.

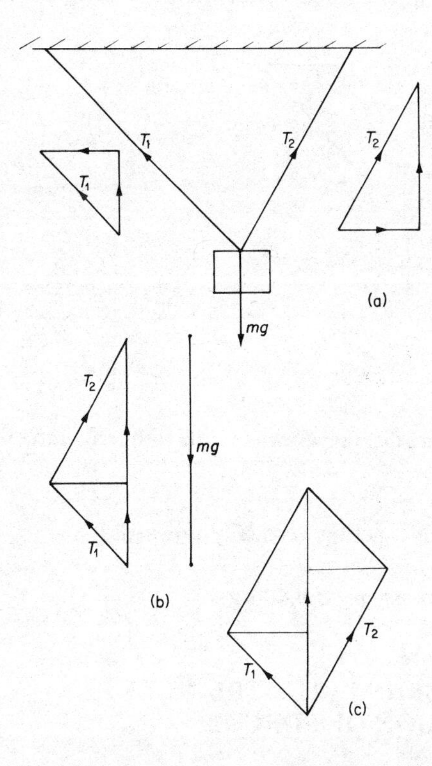

Figure 2.4 (a) The mass m has its gravitational attraction mg supported by the tensions T_1 and T_2 in the wires; the tensions T_1 and T_2 can be split into horizontal and vertical components. (b) The sum of the vertical components is equal and opposite to the gravitational attraction mg; note that the horizontal components are equal and opposite and balance each other out. (c) Vectors representing the tensions T_1 and T_2 arranged as the adjacent sides of a parallelogram

Example 2.1

An unknown mass is supported by two wires A and B. The tension in wire A, inclined at 45° to the horizontal, is 48 N and the tension in wire B, inclined at 60° to the horizontal, is 68 N. Determine the

gravitational force exerted on the mass and hence find the mass.
Solution From the force vector diagram in figure 2.5

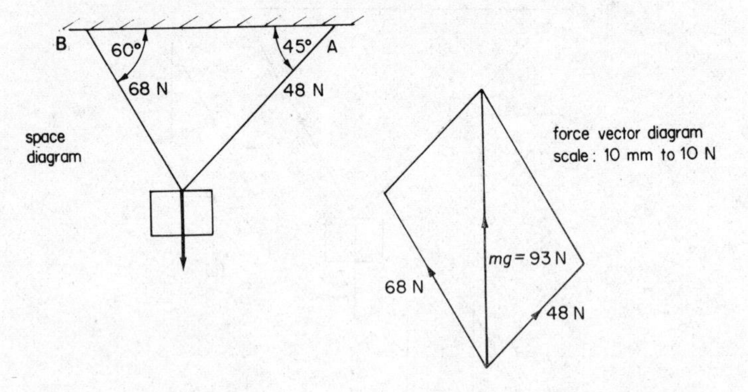

Figure 2.5 Space diagram drawn to scale with vector force diagram drawn to scale

$$mg = 93 \text{ N} \quad \text{acting vertically upwards}$$

therefore the mass, m, is 9.48 kg.

2.2 EQUILIBRIUM—THE TRIANGLE AND POLYGON OF FORCES

Examination of the parallelogram in figure 2.4 will reveal that the vertical diagonal of the parallelogram (the equilibrant of the two tensions in the wires) can be constructed by adding the two force vectors of the wires. This addition is carried out on each side of the parallelogram, the force vector for each of the tensions being added on to the other force vector, with the diagonal completing the triangle (figure 2.6). Note that the arrows indicating the directions of the forces acting at the point under equilibrium follow each other around the triangle.

This leads us to the theorem of the triangle of forces, which states that if three forces are acting at a point and are in equilibrium, then the force vector diagram will be a triangle with the directions of the forces following each other around the triangle.

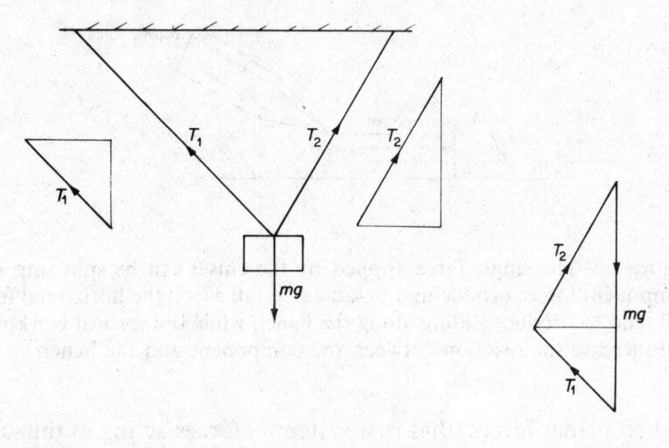

Figure 2.6

This theorem may be used to determine the magnitude and direction of the equilibrant of two forces acting at a point, the direction of the resultant of the two forces being determined by simply reversing the direction of the equilibrant.

Example 2.2

The vertically acting cutting force on a parting-off tool is 1200 N and the horizontally acting force resisting the feeding action of the tool is 400 N. Determine the magnitude and direction of the resultant force acting on the tool.
Solution From figure 2.7

magnitude of resultant = 1250 N
direction of resultant = 72° to horizontal,
acting downwards

We can extend the application of this condition of equilibrium to systems of forces consisting of more than two forces by vectorially adding the forces to form a polygon of forces.

Consider a point subjected to four forces applied to it as shown in figure 2.8. The force vector diagram may be constructed as shown, by drawing the force vectors in sequence working in a clockwise direction around the point. Since the diagram is not a

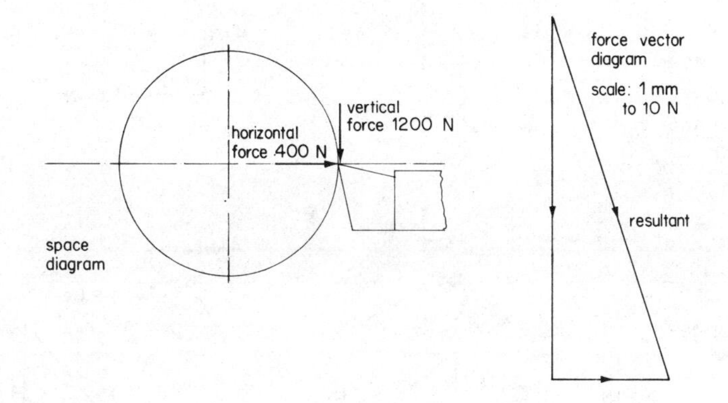

Figure 2.7 Note that the direction of the resultant 'contradicts' the direction of the applied forces

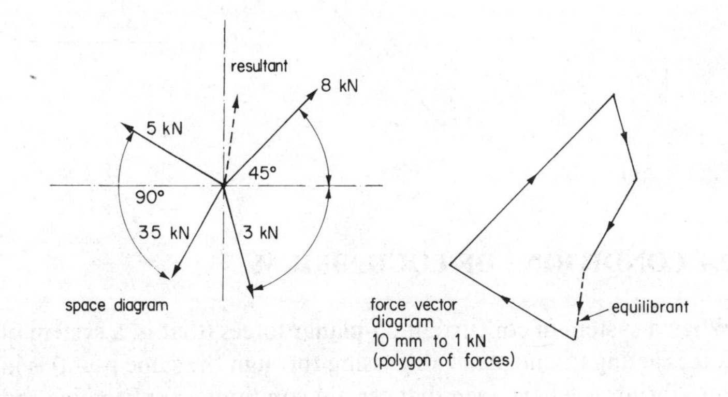

Figure 2.8

closed figure, the point is not in equilibrium but the addition of a force whose vector is the closing (dashed) line of the polygon will produce equilibrium. A single force of equal magnitude but opposite in direction to the equilibrant force, acting through the point, will have the same effect as the four applied forces and will be the resultant of the four applied forces.

Note that in the case of the triangle and polygon of forces the force vector diagram is drawn separately from the space diagram, whereas with the parallelogram of forces the force vector diagram may be superimposed on the space diagram. In all cases it is important to specify the scale being used, stating for example '10 mm to 10 kN' (*note* 10 mm does not *equal* 10 kN).

It should also be noted that the foregoing theorems of parallelograms, triangles and polygons of forces may be applied to vectors representing quantities other than force, for example, the vectors representing velocities (section 5.4).

2.3 BOW'S NOTATION

A difficulty often experienced by many students is in correctly relating a force vector diagram to its corresponding space diagram and vice versa. This can be overcome by using a convenient system of labelling forces known as Bow's notation.

To apply this technique the spaces between the forces in the space diagram, which is usually drawn to scale, are each allocated a distinguishing symbol, usually a letter, working if possible in sequence around the diagram (figure 2.9).

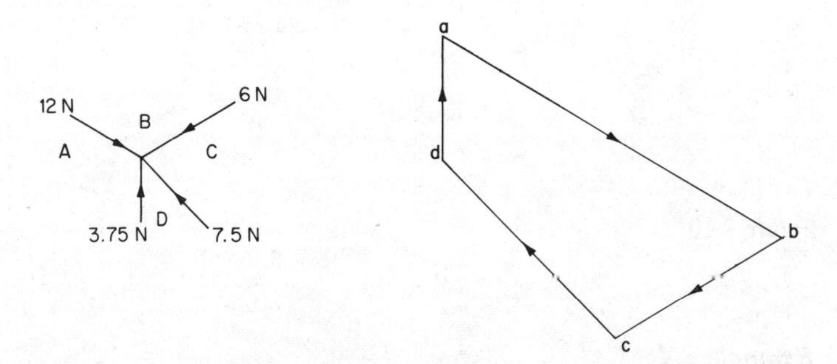

Figure 2.9 Four forces, whose magnitudes and directions are known, acting at a point may be replaced by a single resultant force which will have the same effect; this resultant is equal in magnitude, but opposite in direction, to the equilibrant force, which is found by completing the polygon of forces

Each force is then referred to by the letters appearing on either side of it in the space diagram, again working in clockwise sequence around each point. On the force diagram the vectors, denoted by the letters occurring on either side of them on the space diagram, are added together, with the letters following in sequence around the diagram. This system has the advantage that not only is it possible to relate the space diagram and vector diagram but the sequence of the lettering also assists in deciding on the direction of any unknown forces.

Example 2.3

Determine the resultant of the three-force system shown in figure 2.10.

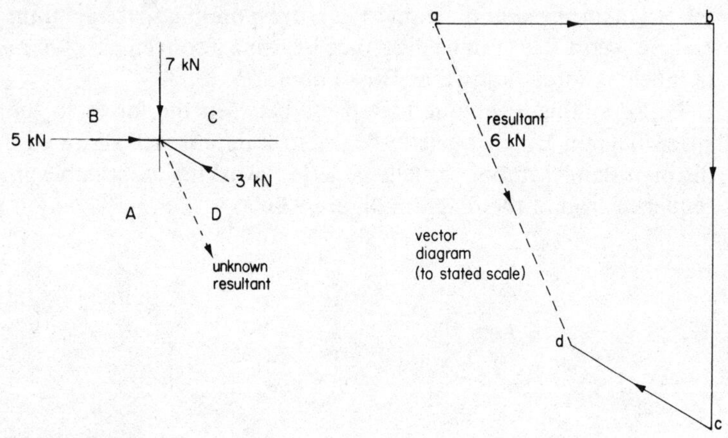

Figure 2.10

Example 2.4

Label the space diagrams in figure 2.11 according to Bow's notation and, by drawing correctly labelled force vector diagrams, determine the unknown forces.

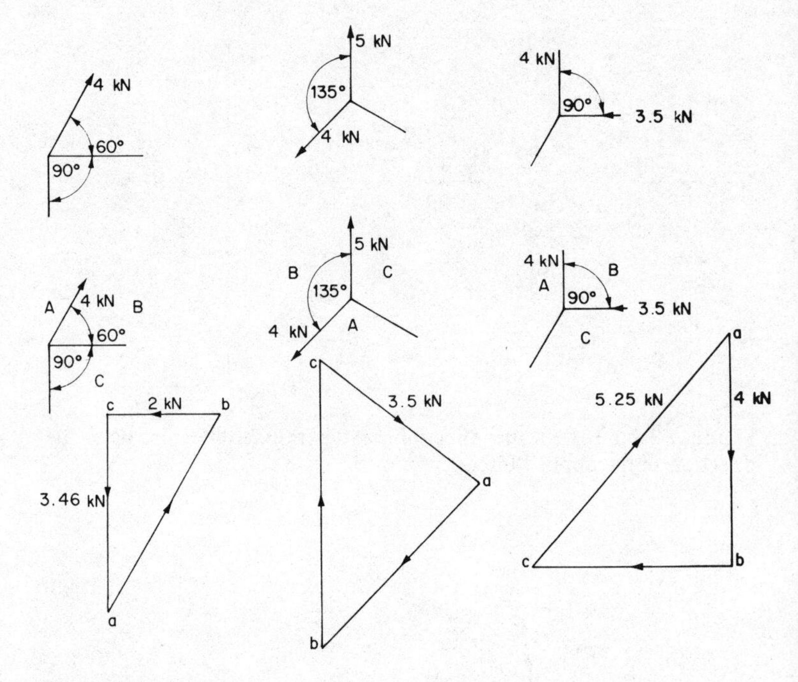

Figure 2.11

2.4 CONDITIONS OF EQUILIBRIUM

When a system of concurrent co-planar forces (that is, a system of forces acting in one plane all passing through the same point) is in equilibrium we have seen that certain conditions apply which can have many useful practical applications. These conditions may be summarised as follows.

(1) If a system of such forces is represented by vectors, adding the vectors in sequence by drawing the 'heads' and 'tails' of the vectors joining each other, the result will be a closed figure. The arrows indicating the directions of the forces will follow each other all round the diagram. This condition is a development of the triangle of forces; we used this method in the graphical determination of resultants and equilibrants and we shall be applying it in the solution of force in frameworks.

(2) A consequence of this first condition is that if all the forces are split into two components acting in two mutually perpendicular planes (usually horizontal and vertical) then the algebraic sum of the components in these two directions will be zero. 'Algebraic sum' means adding the vectors taking into account the directions in which they are acting. This condition is also used to determine the resultant forces in frameworks.

(3) The third condition derives from the fact that there is no tendency to rotate when equilibrium exists. This is expressed mathematically by stating that when equilibrium exists the algebraic sum of the turning moments about any point is equal to zero. An alternative method of expressing this condition is to say that the clockwise turning moment will equal the anticlockwise turning moments (see section 2.5). This condition of equilibrium is applied to the problems of determining the centres of gravity of solids, the centres of area of sections and laminae, and the support reactions on beam and framework problems.

Example 2.5

Determine graphically and by resolution the magnitude and direction of the equilibrant force of the three-force system shown in figure 2.12.

Solution Graphically—see polygon of force vector diagram, figure 2.12.

By resolution: each force is split into its horizontal and vertical components, as shown in figure 2.12. The vertical components are algebraically added, taking forces acting upwards as positive. Thus

$$4 \sin 30 + 3 \sin 45 - 6 \sin 60 + V = 0$$
$$2 + 2.12 - 5.196 + V = 0$$
$$V = 1.076 \text{ N}$$

The horizontal components are algebraically added, taking components acting to the right as positive. Thus

$$3 \cos 45 - 4 \cos 30 - 6 \cos 60 + H = 0$$
$$2.12 - 3.464 - 3 + H = 0$$
$$H = 4.344 \text{ N}$$

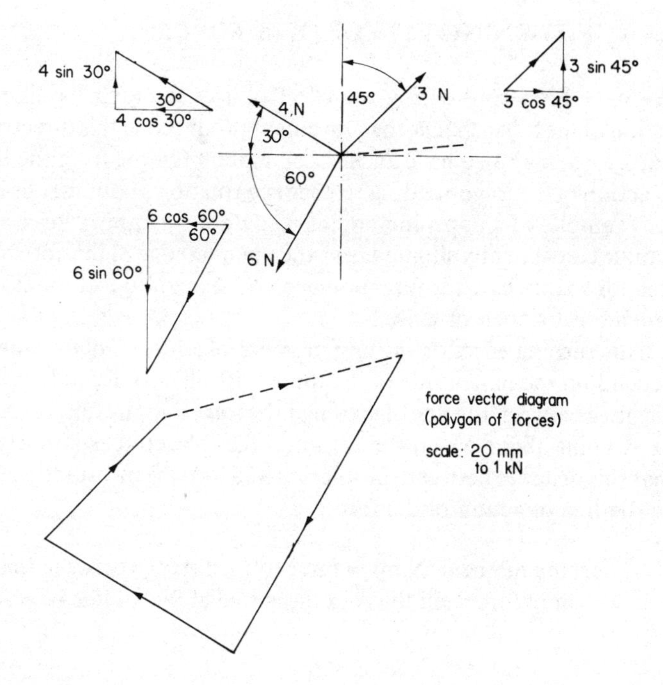

Figure 2.12

The resultant force, R, is given by

$$R^2 = V^2 + H^2$$

thus

$$R = \sqrt{(1.076^2 + 4.344^2)}$$
$$= \sqrt{(1.158 + 18.87)}$$
$$= 4.48 \text{ N}$$

Resultant direction, θ, is given by

$$\tan \theta = \frac{1.076}{4.344} = 0.25$$

$$\theta = 12°9'$$

2.5 THE TURNING EFFECT OF A FORCE

We have been considering the effect of a force along the line of action of the force, that is, the movement produced in this direction, but a force may also have an effect on a point remote from the line of action of the force, tending to produce rotation about that point; for example, when opening or closing a door we apply a force at a handle conveniently situated near the opening edge of the door and this force applies a turning moment about the hinge of the door, causing it to open or close.

This turning effect or turning moment of a force will obviously depend on the magnitude of the force but will also depend on the distance between the line of action of the force and the fulcrum (the fixed point about which the turning takes place). It is important that this distance between the fulcrum and force be measured at 90° to the line of action of the force.

Turning moment (N m) = force (N) × distance between line of action of force and fulcrum measured at 90° to line of action (m)

It is for this reason of obtaining a maximum turning moment with the minimum force, that the door handle is placed as far away from the hinge as possible.

Example 2.6

A force of 15 N is applied at 90° to the axis of a brake lever 300 mm long. Determine the turning moment exerted.

Solution

$$\text{Turning moment} = \text{force} \times \text{distance}$$
$$= 15 \times 0.3 = 4.5 \text{ N m}$$

A bell-crank lever can be used as a device for increasing or decreasing the effort exerted while the turning moment remains constant. Consider the bell-crank lever shown in figure 2.13;

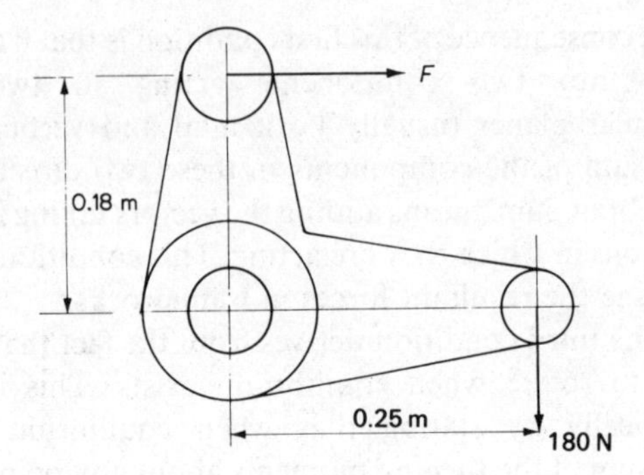

Figure 2.13

equating the turning moment exerted on each arm we have

$$180 \times 0.25 = F \times 0.180$$
$$250 \text{ N} = F$$

Thus a force of 250 N may be exerted in the mechanism while the input effort may only be 180 N, but the 180 N must be moved through a greater distance than the 250 N (see sections 4.1 and 4.2).

The same principle can be extended through several levers in a mechanism to transmit a force to a remote position.

Several forces may each exert a turning moment about the same point and if they tend to produce rotation in the same direction the total turning moment may be found by adding the individual turning moments together. Likewise when a force exerts a turning moment in the opposite direction, this turning moment may be subtracted from those acting in the opposite direction to find the total turning moment.

We have already said (section 2.4) that when a point is in equilibrium there is no resultant turning moment and consequently in this case the algebraic sum of turning moments is equal to zero.

Example 2.7

A simple press for inserting ball-bearings in a gearbox consists of a lever 0.6 m long pivoted at one end. The force to the bearing is applied at a point 0.1 m from the pivot. If the effort at the end of the lever is 40 N, determine the force applied to the bearing.

Solution

$$\text{Moment due to applied effort} = \text{moment due to force on bearing}$$

$$0.6 \times 40 = 0.1 \times F$$

$$24 \text{ N m} = 0.1\ F$$

therefore

$$F = \frac{24}{0.1} = 240 \text{ N}$$

2.6 BEAM REACTIONS

One of the most common applications of this condition of equilibrium is that of beam reactions. Consider a horizontal beam simply supported at two points along its length and carrying loads which can be imagined to be concentrated at points along its length. A practical application of this problem may be found in beams in buildings, bridges and beds of machine tools. If the loads are imagined to be acting vertically then the reactions at the supports may also be taken to be acting vertically, as shown in figure 2.14. The directions of these reactions are known but not their magnitudes.

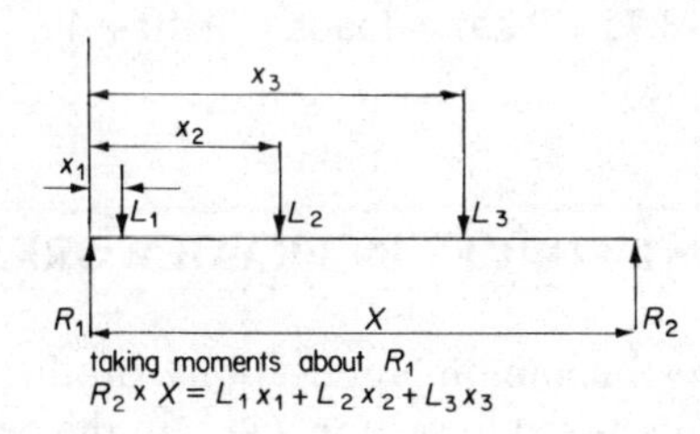

Figure 2.14

Since the beam is in equilibrium, the algebraic sum of the turning moments about any point is zero, but two of the moments will be due to the unknown reactions. However, if the point about which the moments are taken is made to coincide with one of the reactions, then the turning moment due to that reaction will be zero (since the distance by which the reaction should be multiplied to obtain the turning moment is zero) and only one quantity will remain unknown.

Referring to figure 2.14 and taking (that is, calculating) moments about R_1, the turning moments due to the loads will be clockwise and the total clockwise turning moment will be resisted by the anticlockwise moment due to R_2. This enables R_2 to be determined. Then by taking moments about the reaction R_2, the unknown reaction R_1 may be calculated.

Finally by applying another condition of equilibrium we may check the accuracy of our results. Since the whole assembly is in equilibrium the downward acting loads L_1, L_2 and L_3 must be equal to the combined upward acting reactions R_1 and R_2. Thus

$$L_1 + L_2 + L_3 = R_1 + R_2$$

Example 2.8

A beam 2.5 m long, simply supported at its ends, carries vertically acting point loads as follows: 3 kN at 0.75 m from the left-hand support and 4 kN at 1.5 m from the left-hand support. Determine the reactions set up at the supports.

Solution See figure 2.15. Taking moments about R_1 (the left-hand support)

$$\text{anticlockwise moments} = \text{clockwise moments}$$

$$R_2 \times 2.5 = (3 \times 0.75) + (4 \times 1.5)$$

$$= 2.25 + 6 = 8.25$$

therefore

$$R_2 = \frac{8.25}{2.5} = 3.3 \text{ kN}$$

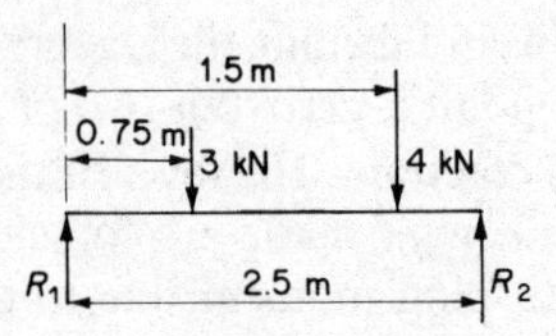

Figure 2.15

Taking moments about R_2 (the right-hand support)

$$\text{clockwise moments} = \text{anticlockwise moments}$$
$$R_1 \times 2.5 = (4 \times 1) + (3 \times 1.75)$$
$$= 4 + 5.25 = 9.25$$

therefore

$$R_1 = \frac{9.25}{2.5} = 3.7 \text{ kN}$$

Check

$$\text{reactions } (3.3 + 3.7) = \text{loads } (3 + 4)$$
$$7 \text{ kN} = 7 \text{ kN}$$

Note By taking moments about each point of support and using the 'loads = reactions' condition of equilibrium as a check, the possibility of mathematical errors will be eliminated.

Example 2.9

A beam of length 4 m is simply supported over a span of 3 m with an overhang as shown in figure 2.16a. If vertically acting point loads are applied as shown, determine the reactions R_1 and R_2.
Solution Taking moments about R_1

$$(2 \times 0) + (7 \times 0.75) + (10 \times 2) + (4 \times 4) = R_2 \times 3$$
$$0 + 5.25 + 20 + 16 = 41.25 = R_2 \times 3$$

$$R_2 = \frac{41.25}{3} = 13.75 \text{ N}$$

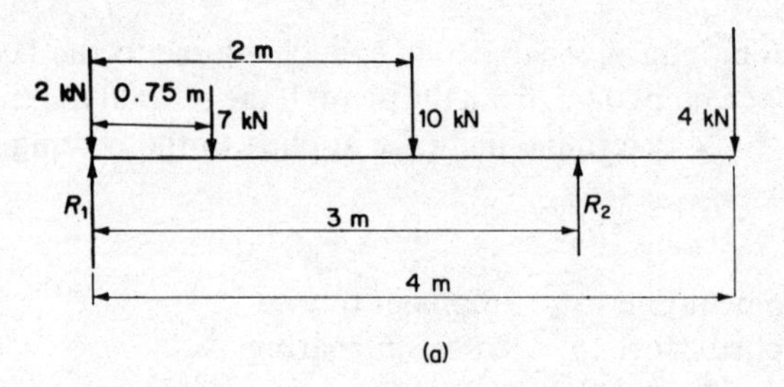

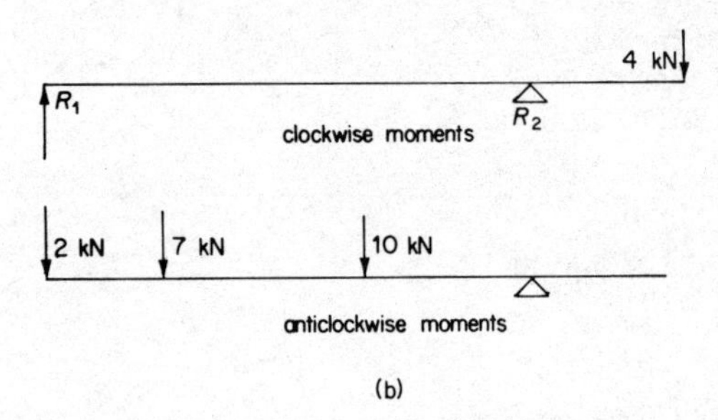

Figure 2.16

Taking moments about R_1 (see figure 2.16b)

$$(4 \times 1) + (R_1 \times 3) = (10 \times 1) + (7 \times 2.25) + (2 \times 3)$$
$$= 10 + 15.75 + 6 = 31.75$$

$$R_1 = \frac{31.75 - 4}{3} = 9.25 \text{ N}$$

Check

$$\text{reactions } (13.75 + 9.25) = \text{loads } (7 + 10 + 4 + 2)$$
$$23 \text{ kN} = 23 \text{ kN}$$

2.7 RESULTANT FORCES IN FRAMEWORKS

One of the most common applications of the conditions of equilibrium, as discussed in section 2.4, is in the determination of the forces set up in the members of a framework such as might be met in roof structures, cranes and bridges as the result of external loads.

However, in applying these conditions we make certain assumptions that may not be true in practice. Firstly we assume that all the joints are pin-jointed (and not rivetted or welded as occurs in practice). This will ensure that the internal forces set up in the members as a result of the external loads will act axially along the members and be entirely tensile or compressive; then we know the lines of application of each of the forces in the members. Secondly we assume that the loads applied to the framework are concentrated at and act through the joints.

These assumptions are applied in the techniques demonstrated in the following sections. Furthermore it should be noted that the forces set up in a loaded framework are unaffected by the size of the framework.

2.8 GRAPHICAL DETERMINATION OF FORCES IN SIMPLY SUPPORTED FRAMEWORKS SUBJECT TO VERTICAL LOADS

In this particular type of problem the positions of both of the reactions are known and also their direction—they both act vertically since the loads applied only act vertically.

The first step is to determine the magnitude of the reactions by applying the condition of equilibrium, which states that for equilibrium the algebraic sum of the turning moments about any point is zero.

The lines of action of the forces are extended and the problem may be treated as a beam. By taking moments about each point of support in turn the reactions are calculated. The values may be checked by equating the sum of the reactions to the total load applied.

The next stage is to construct the force vector diagram for the external loads and supporting reaction.

These external vertically acting loads and reactions may be represented graphically by means of a straight vertical line whose total length is the sum of the loads (or of the reactions), each load and reaction being added working in sequence, say clockwise, around the framework.

The next step is to consider each joint in turn. Each joint is in equilibrium and consequently each joint will have a fully closed force diagram which is constructed by adding in sequence the forces acting around the joint. These individual force diagrams may be superimposed on each other and on to the external force diagram (vertical line referred to above). Information obtained from one joint is often used to help construct the force diagram for an adjacent joint.

Since each member of the framework is in equilibrium the internal force set up at one end of the member as the result of equilibrium at one joint must be balanced by an equal and opposite force acting at the other end of the member.

The direction of the internal resisting forces acting in each member should be indicated on the space diagram adjacent to the joint. The members with the internal forces acting outwards, against the external inward acting forces, are in compression and are known as *struts*. Those members whose internal forces are pulling inwards are under tension from external forces and are known as *ties*.

Example 2.10

Determine the forces set up in the members of the simply supported framework subjected to vertically acting loads shown in figure 2.17.

Solution Taking moments about R_2 to find R_1

$$\begin{aligned} R_1 \times 16 &= (5 \times 8) + (5 \times 4) \\ &= 40 + 20 \end{aligned}$$

$$R_1 = \frac{60}{16} = 3.75 \text{ kN}$$

Taking moments about R_1 to find R_2

$$\begin{aligned} R_2 \times 16 &= (5 \times 8) + (5 \times 12) \\ &= 40 + 60 \end{aligned}$$

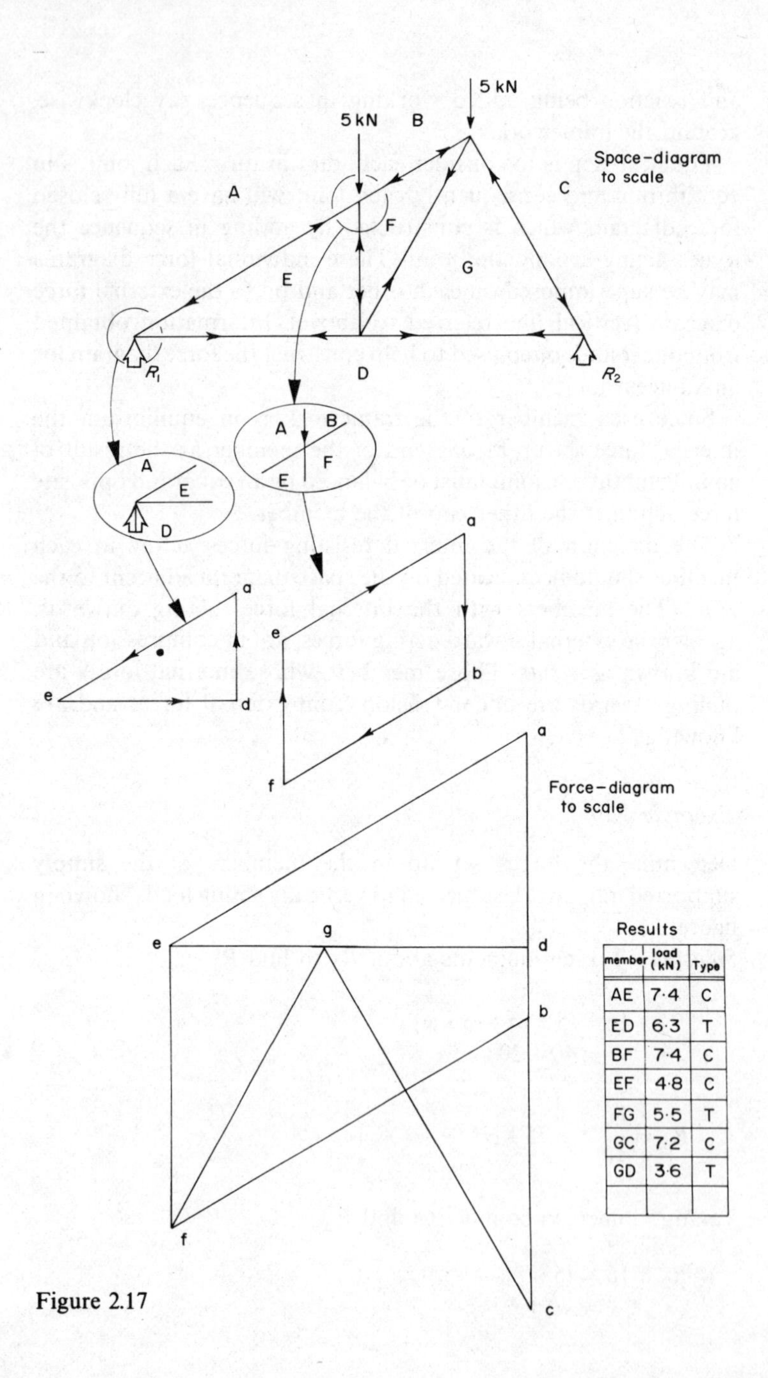

member	load (kN)	Type
AE	7·4	C
ED	6·3	T
BF	7·4	C
EF	4·8	C
FG	5·5	T
GC	7·2	C
GD	3·6	T

Figure 2.17

$$R_2 = \frac{100}{16} = 6.25 \text{ kN}$$

Check

reactions $(3.75 + 6.25)$ = loads $(5 + 5)$
10 kN = 10 kN

2.9 GRAPHICAL DETERMINATION OF FORCES IN SIMPLY SUPPORTED FRAMEWORKS SUBJECT TO VERTICAL AND NON-VERTICAL LOADS

In many practical cases a framework will not only be subjected to vertically acting loads, due possibly to its own mass or the roof cladding, but also due to forces acting in directions other than vertical, due possibly to wind forces. Such an example is shown in figure 2.18.

A framework so loaded will exert a non-vertical reaction at the supports so that it may be in equilibrium under the action of the external loads and supports.

This is accomplished by arranging for one of the joints to act as a hinge, which will accommodate the non-vertical reaction, and for the other joint to be supported on rollers, usually moving on a horizontal surface. These rollers have the advantage that they will accommodate small dimensional changes, due to temperature fluctuations, and the reactions will act normal to the supporting face. Thus, while the direction of the reaction at the rollers will be known, the direction of the reaction through the hinge will not be known.

However, if we take moments about the hinge, the moment of this totally unknown reaction will be zero and the magnitude of the other reaction (at the rollers) can be determined. This reaction is then known in magnitude and direction, enabling the vector diagram for all the external loads and reactions to be drawn. The vectors for the loads and vertical reaction are plotted, working in sequence around the framework (figure 2.18b). By closing this external force diagram, the magnitude and direction of the reaction acting through the hinge can be determined.

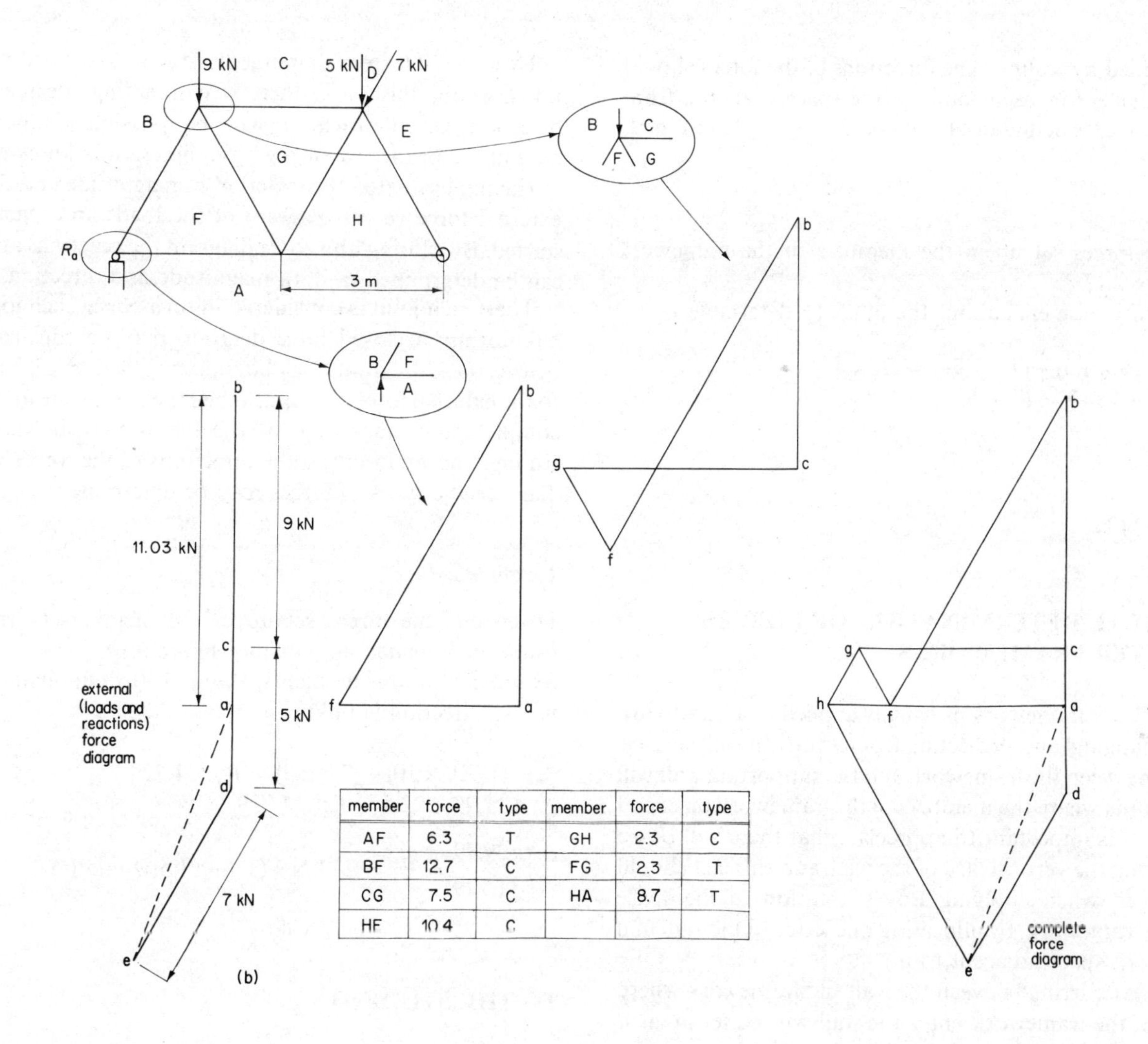

member	force	type	member	force	type
AF	6.3	T	GH	2.3	C
BF	12.7	C	FG	2.3	T
CG	7.5	C	HA	8.7	T
HF	10.4	C			

Figure 2.18

The internal forces set up within the members of the framework can be found by considering the force diagram for each joint in turn. As before, the force diagram for each joint will be a closed figure and information from one force diagram can be used to assist the construction of other diagrams. These diagrams may be superimposed on each other and on to the external force diagram to obtain a composite force diagram for the framework as a whole, from which the magnitudes of the forces in each of the members may be determined by scaling. The directions of the forces should be indicated adjacent to each joint on the space diagram, from which the type of force acting in each member may be determined.

Example 2.11

Determine the forces set up in the members of the framework shown in figure 2.17.

Solution Taking moments about the hinge to determine R_a

$$(9 \times 4.5)+(5 \times 1.5)+(7 \times 2.59) = R_a \times 6$$
$$40.5 + 7.5 + 18.13 = R_a \times 6$$
$$\frac{66.13}{6} = R_a$$
$$R_a = 11.02 \text{ kN}$$

2.10 GRAPHICAL DETERMINATION OF FORCES IN CANTILEVER FRAMEWORKS

The term 'cantilever framework' is usually applied to a framework which is 'overhanging' and projecting from a vertical wall surface. The reactions between the framework and the supporting wall will occur at the points where the members of the framework meet and enter the wall. It is important to appreciate that there will be no force acting along the vertical face of the wall, and this fact should be acknowledged when applying Bow's notation to the space diagram of the framework, by allocating one letter to the wall and to the framework space adjacent to it.

The reactions occurring between the wall and framework where the members of the framework enter the wall will be set up as a result of the forces in those members of the framework due to the applied loads. Consequently where only one member enters the wall (R_1 in figure 2.19) the reaction will be in line with that member, but where a joint between the wall and the framework involves two or more members of the framework then the actual direction of the reaction will not be known since it will be the equilibrant (section 2.1) of the forces acting in those members.

However, if (external) moments (due to loads and reactions) are taken about this joint, thereby eliminating the reaction whose direction is unknown, it will be possible to determine the magnitude of the reaction whose direction is known.

Having evaluated this reaction in magnitude and direction, the external force vector diagram of the loads and reactions can be started. By 'closing' this force diagram the reactions at other joints can be determined in both magnitude and direction.

Then each joint is considered in turn. Since each joint will be in equilibrium a closed force diagram may be constructed and as before, these diagrams may be superimposed on each other and on to the external force diagram. When the composite force diagram is complete the magnitude of the forces may be determined by scaling, and by inserting the directions of the forces on the space diagram the types of forces may be determined.

Example 2.12

Determine the forces set up in the members of the cantilever framework loaded as shown in figure 2.19.

Solution Taking moments about X (to eliminate the reaction whose direction is unknown)

$$(1.729 \times 10)+(7.5 \times 3) = R_1 \times 1.729$$
$$17.29 + 22.5 = R_1 \times 1.729$$
$$\frac{39.79}{1.729} = R_1 = 23 \text{ kN} \quad \text{(acting horizontally)}$$

TO THE STUDENT

At the end of this chapter you should be able to

(1) state the general conditions for equilibrium

(2) define and determine the resultant and equilibrant force for a system of forces

(3) determine the reactions at the supports for a simply supported horizontal beam carrying point loads

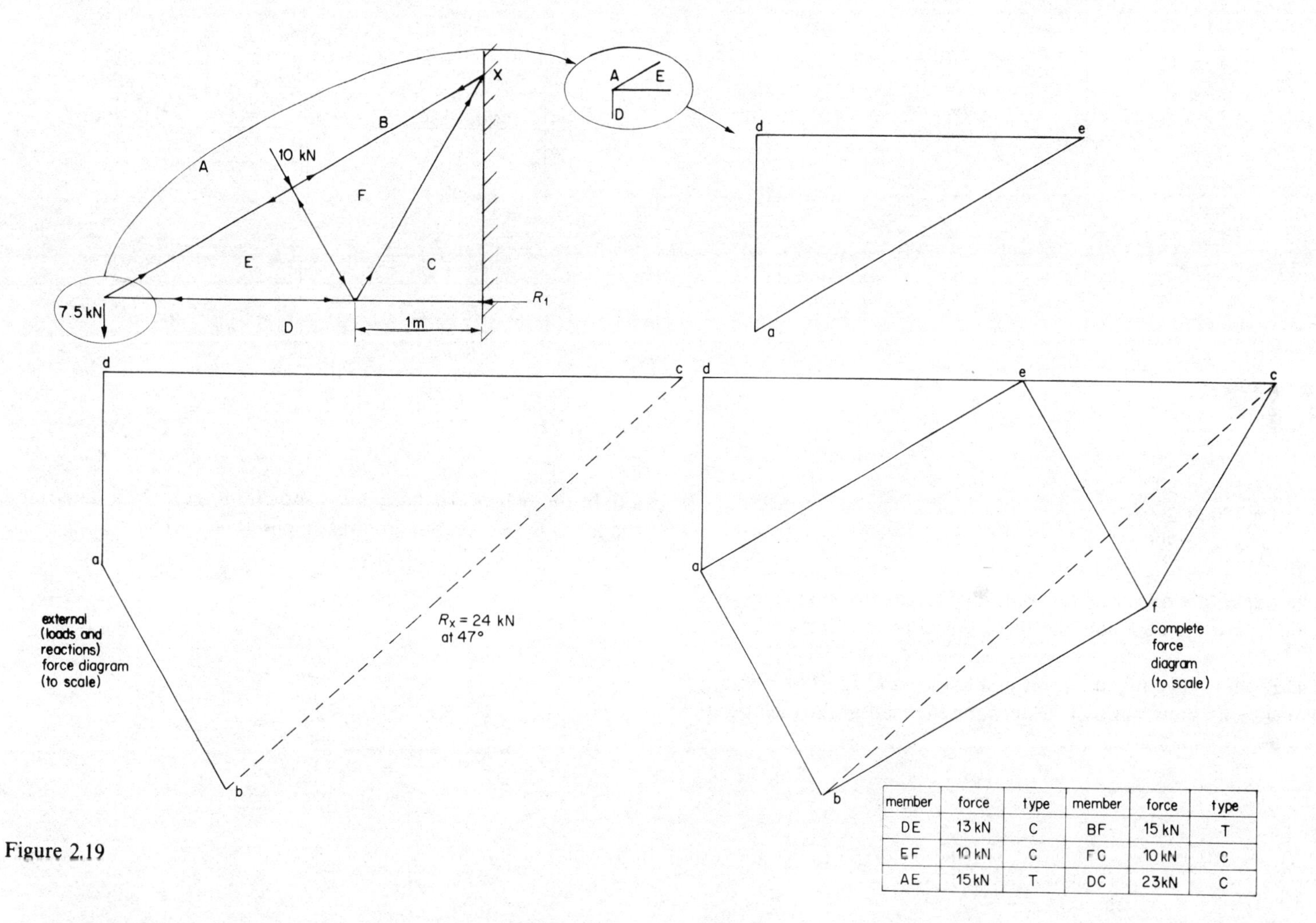

member	force	type	member	force	type
DE	13 kN	C	BF	15 kN	T
EF	10 kN	C	FC	10 kN	C
AE	15 kN	T	DC	23 kN	C

Figure 2.19

(4) determine the nature and magnitude of the resultant forces set up in the members of a loaded framework
(5) complete exercises 2.1 to 2.13.

EXERCISES

2.1 A component of mass 150 kg is supported by two wires both inclined at 30° to the horizontal. Determine the tension in each wire.

2.2 A simple jib crane as shown in figure 2.20 supports a load 1500 N. Determine the forces set up in the two members.

2.3 The following forces are acting at a point with all the forces pulling away from the point.

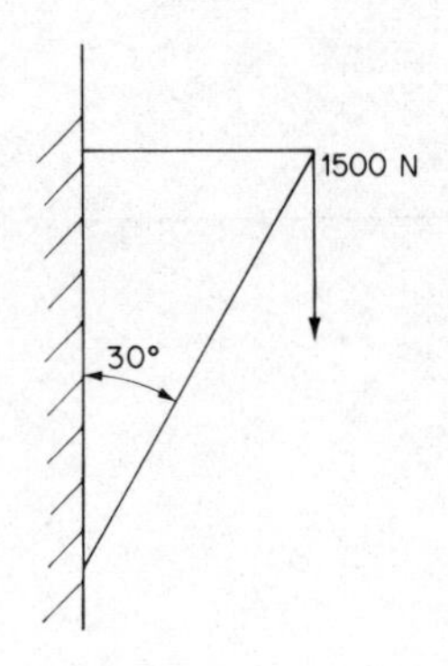

Figure 2.20

10 kN at 30°
6 kN at 90°
1.4 kN at 200°
3 kN at 290°

The angles are all measured from due north working clockwise. Calculate the magnitude and direction of the resultant.

2.4 The forces acting at a joint in a loaded and supported framework are as shown in 2.21. Determine the load in the member P.

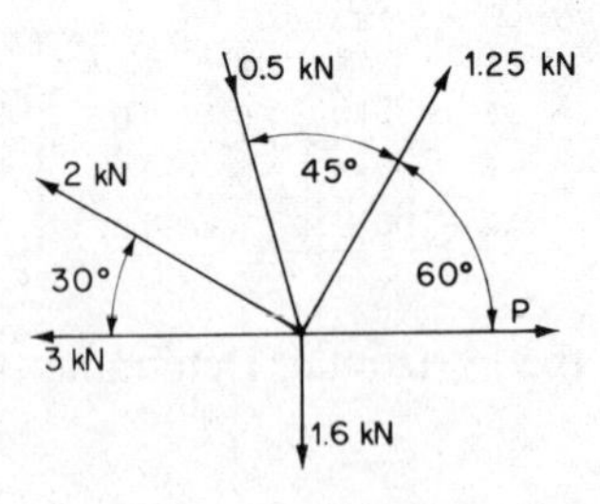

Figure 2.21

2.5 Determine the reactions R_1 and R_2 for the beams loaded as shown in figure 2.22a, b, c and d.

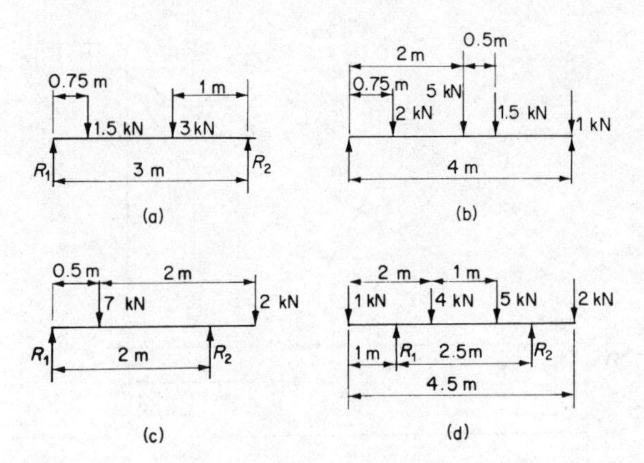

Figure 2.22

2.6 In the bell-crank mechanism shown in figure 2.23, determine the force F to overcome the resistance $R = 2$ MN.

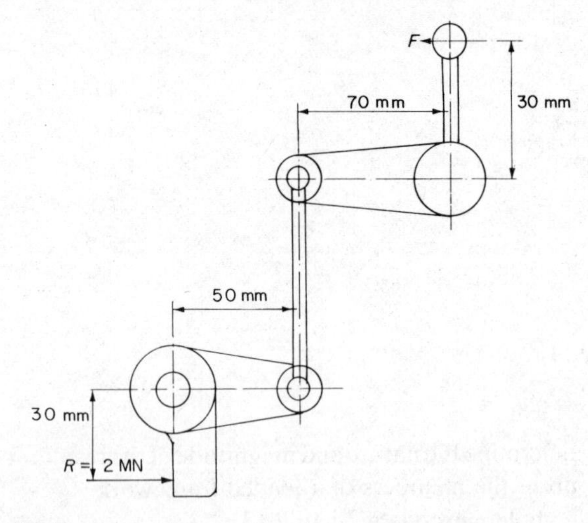

Figure 2.23

2.7 The torque or turning moment required to tighten a nut is 40 N m. If the spanner is 0.375 m long, determine the force required.

2.8 The cutting force on the tool point shown in figure 2.24 is 1500 N, acting vertically downwards. Calculate the force exerted on the securing screw, if the overhang is increased to 45 mm.

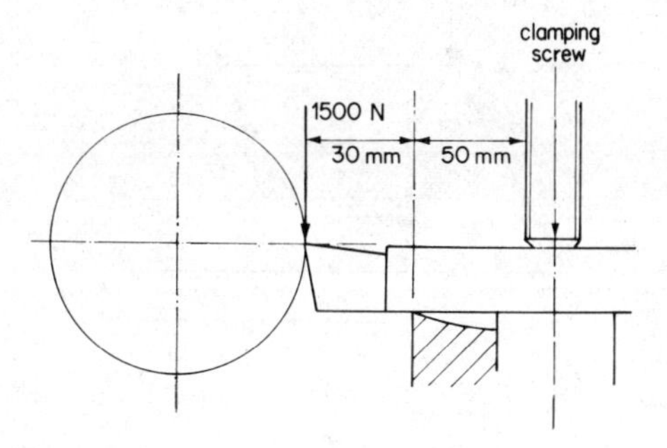

Figure 2.24

2.9 Determine the magnitude and the direction of the resultant for the systems of forces shown in figures 2.25a, b and c. Use a graphical solution in each case and then check your result by calculation.

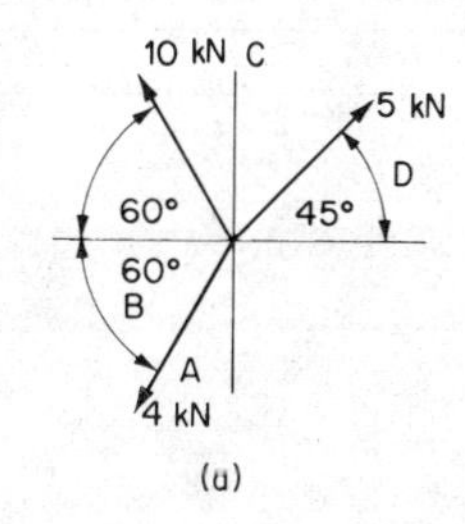

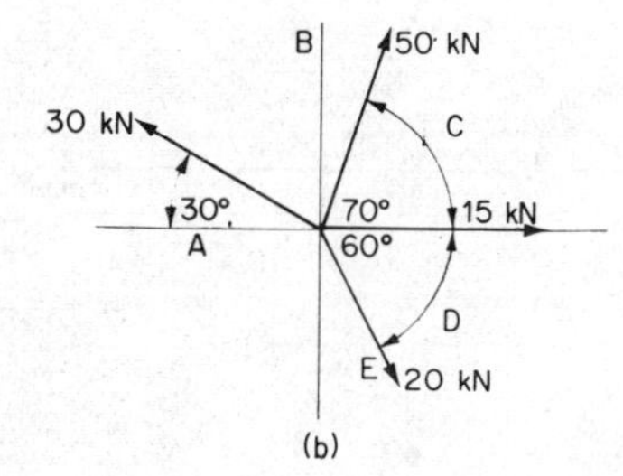

Figure 2.25

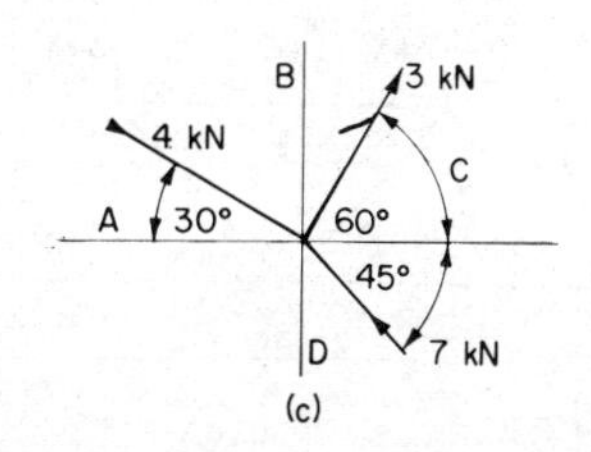

Figure 2.25 (*continued*)

2.10 In the force systems shown in figures 2.26a and b determine the magnitude of the unknown forces acting in the directions CD and AD.

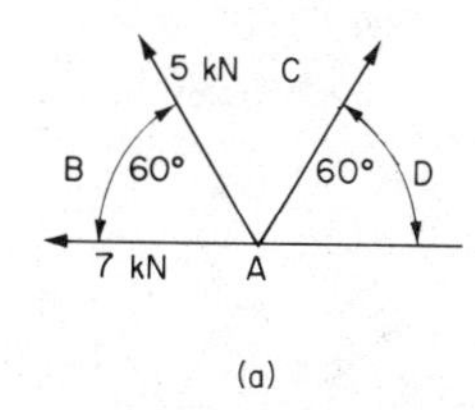

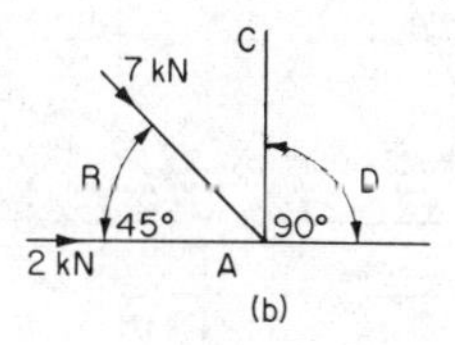

Figure 2.26

2.11 Determine the magnitude and nature of the forces set up in the members of the simply supported frameworks shown in figures 2.27a, b and c as a result of the vertically acting loads.

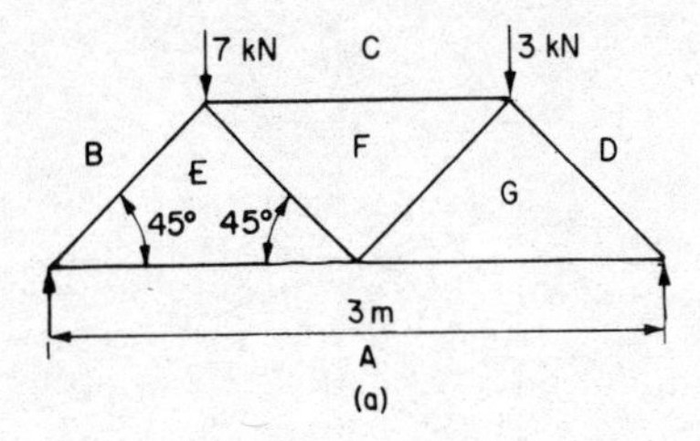

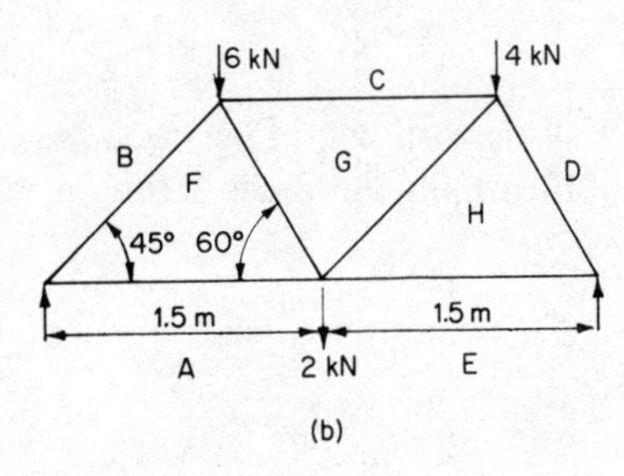

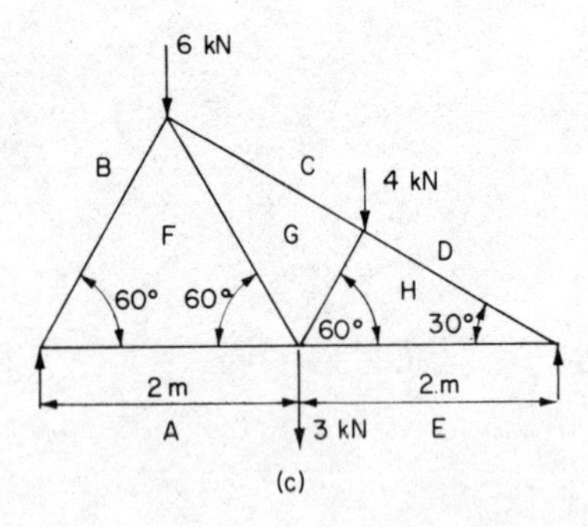

Figure 2.27

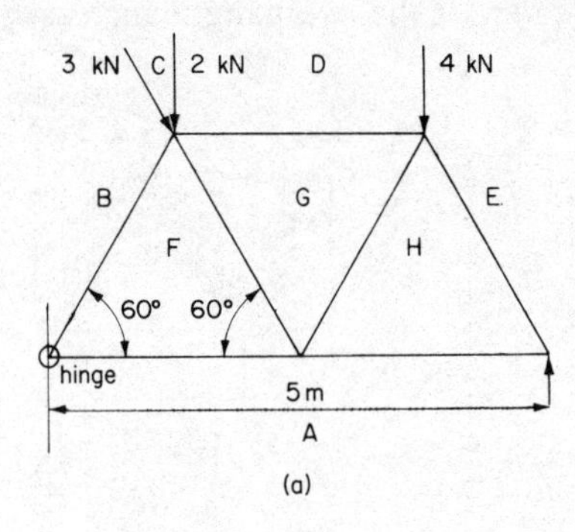

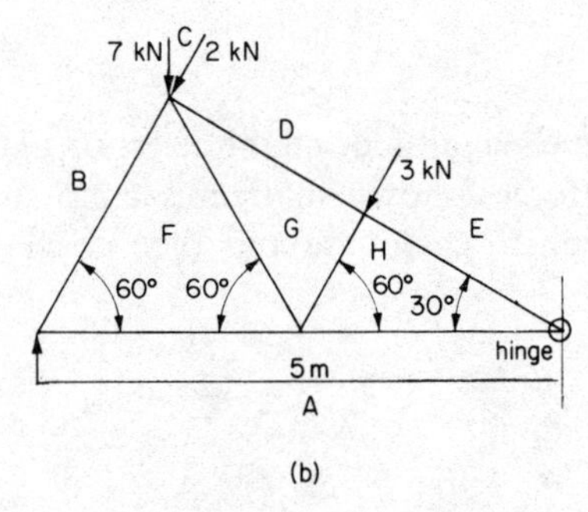

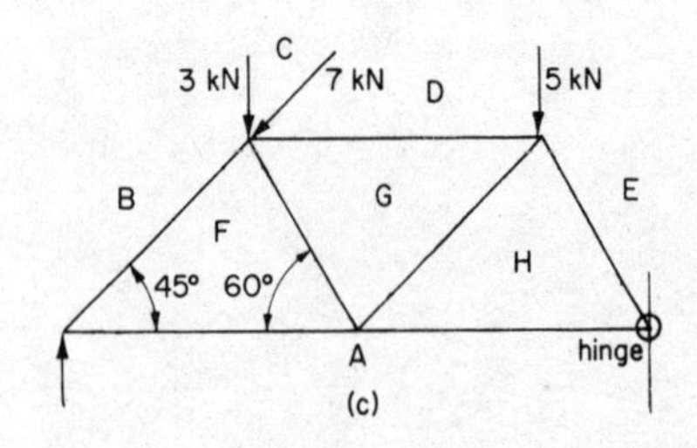

2.12 Determine the magnitude and nature of the forces set up in the members of the frameworks hinged and supported as shown in figures 2.28a, b and c due to the loads shown.

Figure 2.28

2.13 Determine the magnitude and nature of the forces set up in the members of the cantilever frameworks shown in figures 2.29a, b and c as a result of the loads applied as shown.

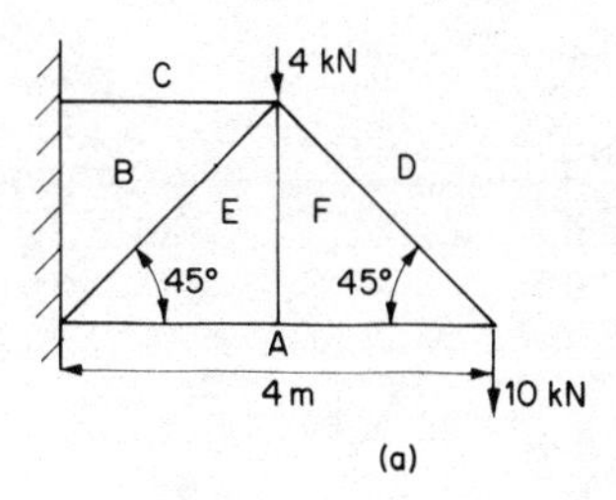

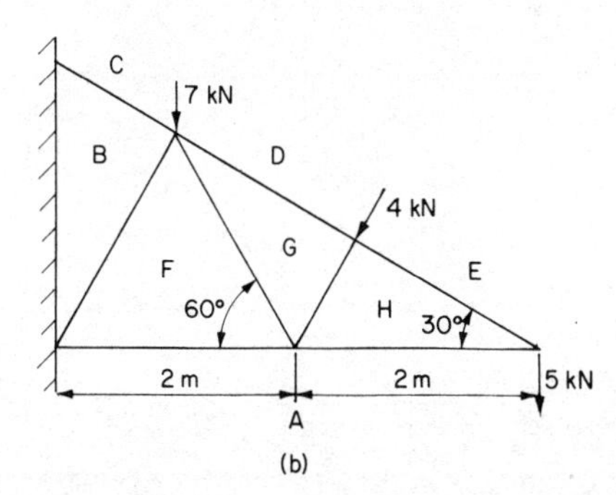

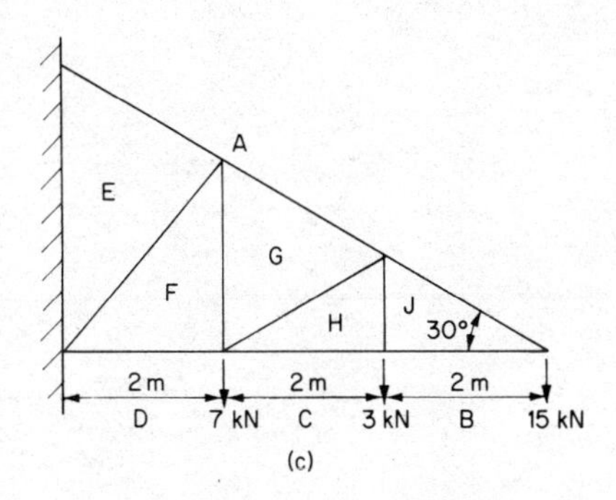

Figure 2.29

NUMERICAL SOLUTIONS

2.1 1471.5 N
2.2 866 N tension; 1732 N compression
2.3 11.4 kN at 43°
2.4 3.9 kN
2.5 (a) $R_1 = 2.125$ kN, $R_2 = 2.375$ kN, (b) $R_1 = 4.688$ kN, $R_2 = 4.812$ kN, (c) $R_1 = 4.75$ kN, $R_2 = 4.25$ kN, (d) $R_1 = 4$ kN, $R_2 = 8$ kN
2.6 2.8 MN
2.7 107 N
2.8 1350 N
2.9 (a) 9.4 kN at 341.5°, (b) 47.5 kN at 200°, (c) 5.5 kN at 180°
2.10 (a) CD = 5 kN, DA = 12 kN, (b) CD = 5 kN, DA = 7 kN
2.11 See figure 2.30

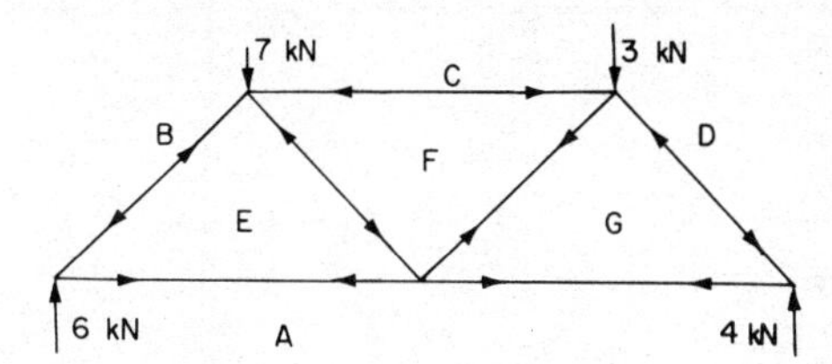

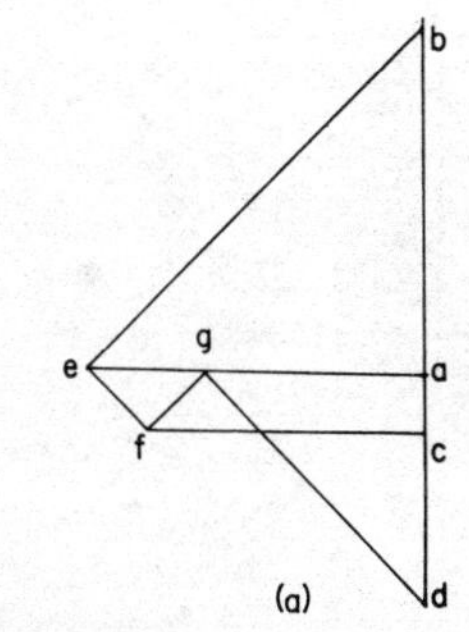

member	force	type
BE	8.5	C
CF	5.0	C
GD	5.5	C
GA	4.0	T
EA	6.0	T
EF	1.5	C
FG	1.5	T

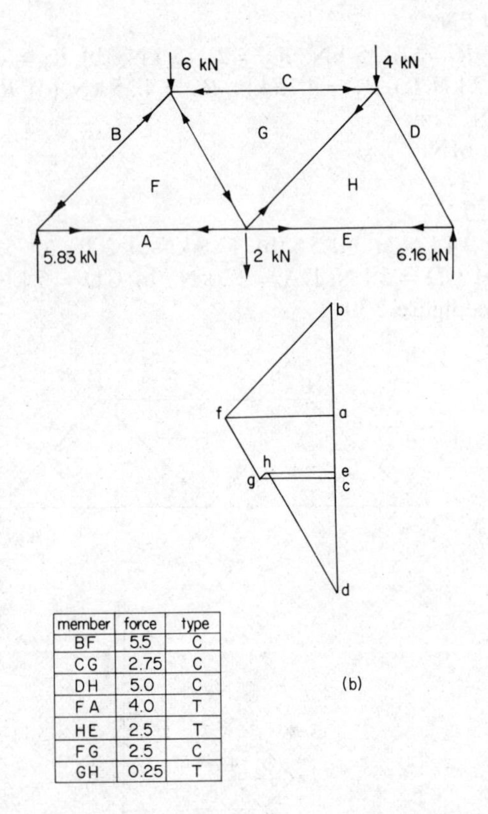

member	force	type
BF	5.5	C
CG	2.75	C
DH	5.0	C
FA	4.0	T
HE	2.5	T
FG	2.5	C
GH	0.25	T

(b)

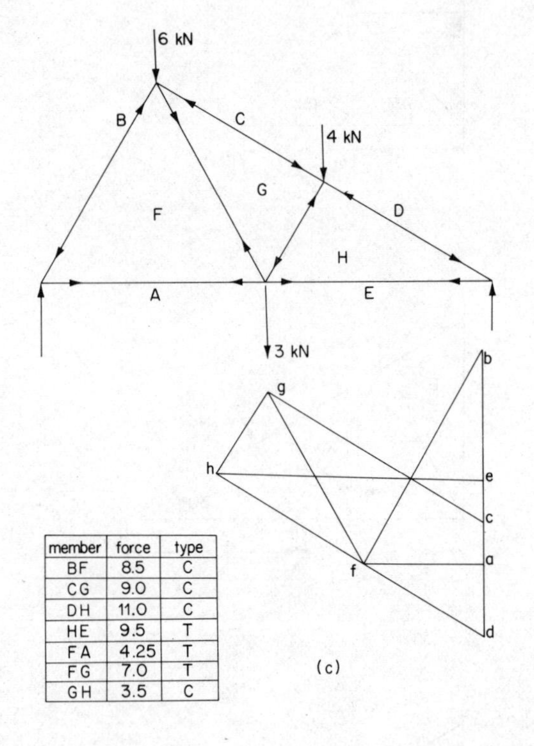

member	force	type
BF	8.5	C
CG	9.0	C
DH	11.0	C
HE	9.5	T
FA	4.25	T
FG	7.0	T
GH	3.5	C

(c)

Figure 2.30

2.12 See figure 2.31

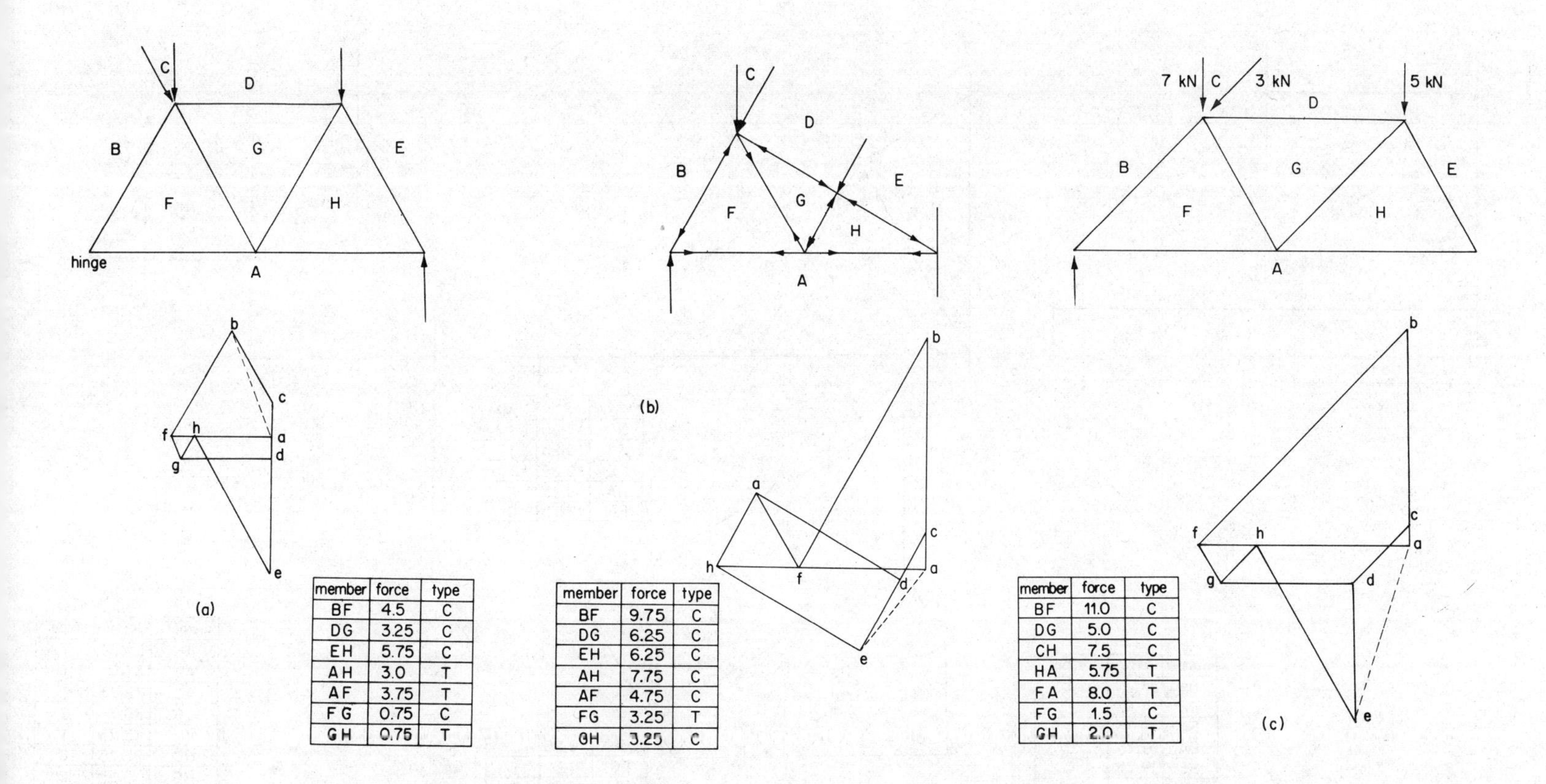

member	force	type
BF	4.5	C
DG	3.25	C
EH	5.75	C
AH	3.0	T
AF	3.75	T
FG	0.75	C
GH	0.75	T

member	force	type
BF	9.75	C
DG	6.25	C
EH	6.25	C
AH	7.75	C
AF	4.75	C
FG	3.25	T
GH	3.25	C

member	force	type
BF	11.0	C
DG	5.0	C
CH	7.5	C
HA	5.75	T
FA	8.0	T
FG	1.5	C
GH	2.0	T

Figure 2.31

2.13 See figure 2.32

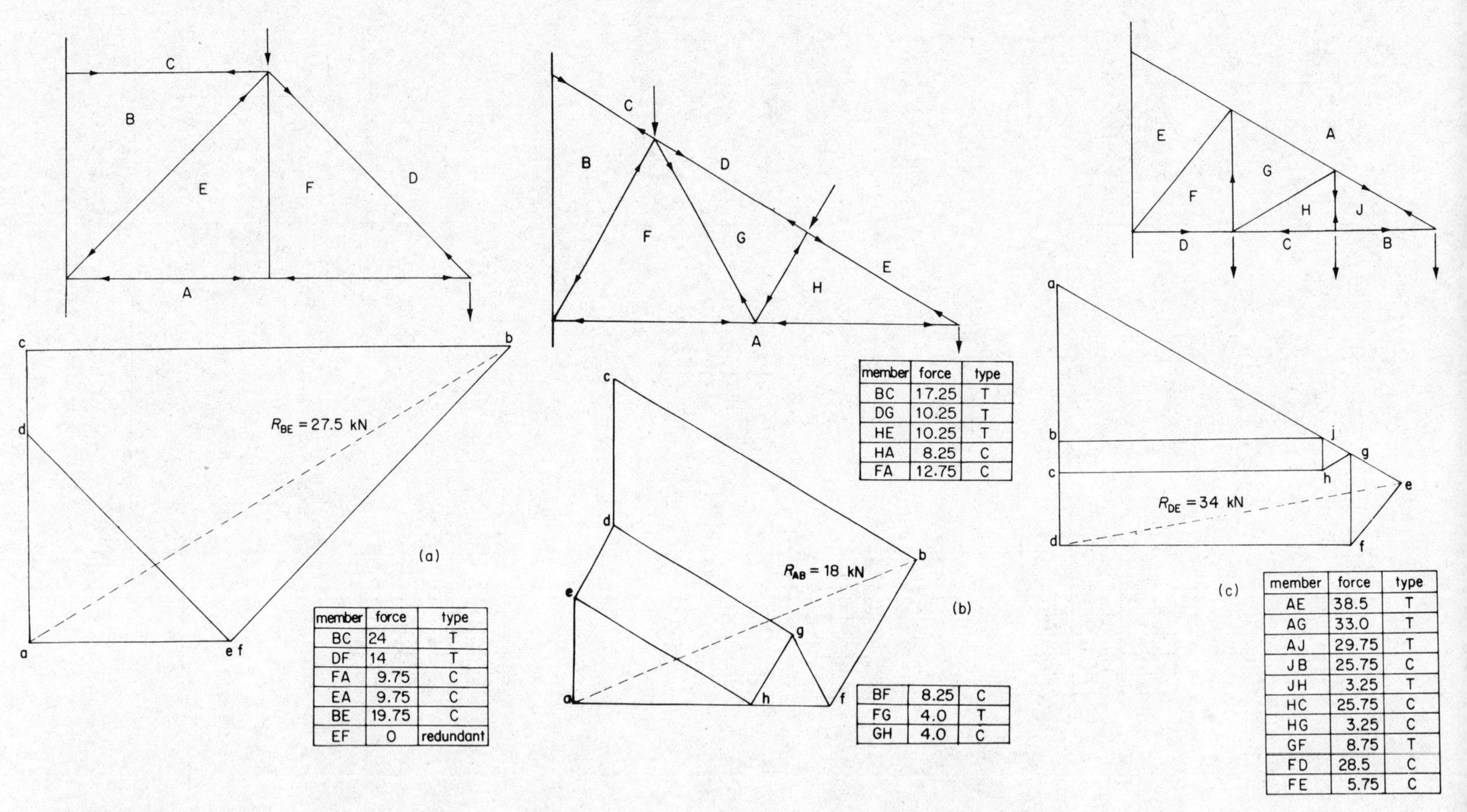

member	force	type
BC	24	T
DF	14	T
FA	9.75	C
EA	9.75	C
BE	19.75	C
EF	0	redundant

member	force	type
BC	17.25	T
DG	10.25	T
HE	10.25	T
HA	8.25	C
FA	12.75	C
BF	8.25	C
FG	4.0	T
GH	4.0	C

member	force	type
AE	38.5	T
AG	33.0	T
AJ	29.75	T
JB	25.75	C
JH	3.25	T
HC	25.75	C
HG	3.25	C
GF	8.75	T
FD	28.5	C
FE	5.75	C

Figure 2.32

3 Dry Sliding Friction

The object of this chapter is that the student shall be introduced to the nature of dry sliding friction and is able to appreciate the effects of dry sliding friction.

3.1 THE NATURE OF DRY SLIDING FRICTION

If two flat surfaces are placed in contact with each other and then one of the surfaces is moved relative to the other, or movement is attempted, then this movement is resisted. This resistance to movement is known as *friction.*

To determine why this frictional resistance is always present it is necessary to study what occurs at the interface of the materials.

No matter how flat or smooth a machined surface may appear to be to the naked eye, if we measure the surface finish using special measuring instruments, such as a Talysurf machine, we can see that the surface consists of a series of peaks and valleys, the heights and depths of which will vary according to the quality of the surface finish and the machining process which has been used (figure 3.1). The quality and condition of the machine tool will also cause undulations on the surface. Unmachined surfaces will, of course, produce far rougher appearances when studied under a microscope or similar instrument.

If any two such surfaces of 'valleys and peaks' are placed in contact, then the peaks of one surface will tend to 'lock' into the valleys of the other surface and actual contact between the surfaces will only occur at localised points. Very high pressures will be created at these points of contact and this can lead to minute cold pressure welds being formed.

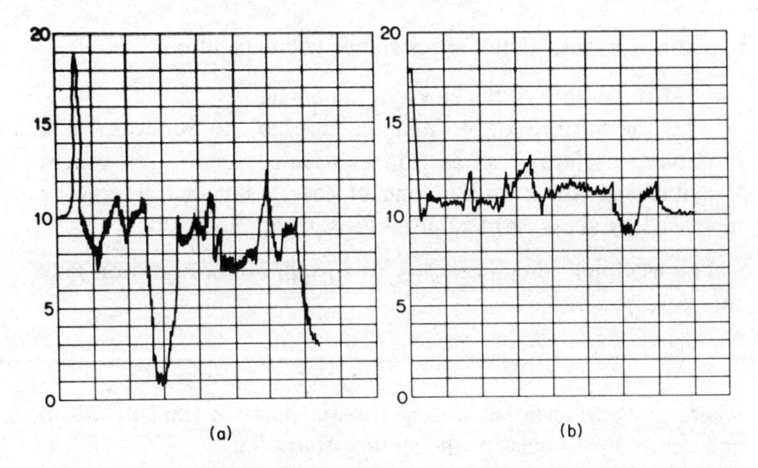

Figure 3.1 Copy traces from a ground steel specimen taken with a Talysurf surface-measuring instrument: (a) transverse profile (across direction of grinding), (b) longitudinal profile

When a force is applied to one surface, attempting to make it slide over the adjacent surface, this interlocking and welding effect will resist motion. This static frictional resistance must be overcome by the applied force before motion can be produced. The magnitude of the static frictional resistance will depend on many factors, including the reaction or force acting between the surfaces, the nature of the surfaces and the time for which they have been in contact.

However, once the applied force has become greater than the frictional resisting force and motion is produced, it is found that the force necessary to sustain the motion, that is, to overcome the dynamic resistance to motion, is slightly less than that required to overcome the static frictional resistance (figure 3.2). It can be shown

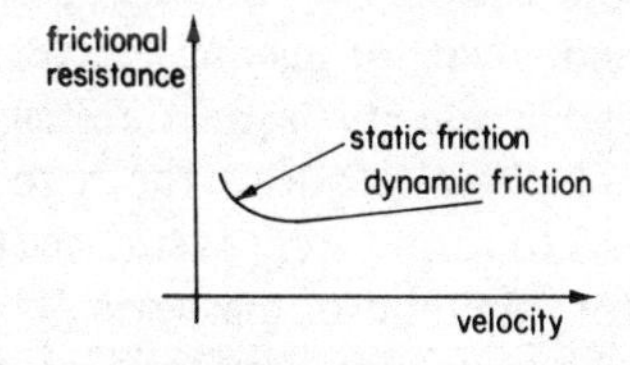

Figure 3.2 Graphs indicating variation of frictional resistance with velocity—the difference between static and dynamic friction

that this dynamic frictional resistance will depend on

(1) the nature of the surfaces in contact
(2) the vertical or normal reaction to the surfaces; on a horizontal surface this reaction is usually due to the weight (gravitational attraction) of one of the masses but it may be increased by some extra external force or applied load.

The frictional force necessary to sustain uniform motion may therefore be written

$$\text{friction force (N)} = \mu mg$$

where μ = coefficient of sliding friction between the surfaces in contact and m = mass of the sliding object (kg).

Furthermore it can be shown that, within limits, the frictional resistance is independent of the areas in contact and independent of the velocity of sliding.

These facts can be expressed in the following so-called *laws of dry sliding friction*, which hold true for most conditions encountered in practice.

(1) The frictional resistance is dependent on the nature of the surfaces in contact.
(2) The frictional resistance is dependent on the force acting at right-angles to the surfaces in contact.
(3) The frictional resistance is independent of the areas in contact.
(4) The frictional resistance is independent of the velocity of sliding.

These laws were expressed by Coulomb,[1] although Leonardo da Vinci had earlier made similar comments, and can be taken as accurate under all but extreme conditions.

They may be varied when, for instance, increasing temperature, possibly due to excessive velocity or pressure, may cause the nature of the surface to alter by breaking down the extremely thin layers of oxide film which tend to form on most surfaces (friction welding).

A simple laboratory method of measuring frictional resistance between two surfaces is to use gravitational effects to move a known mass along a horizontal surface, with the effort being provided by the gravitational attraction on a load connected to the mass by a cord running over a pulley (figure 3.3). When the mass and the friction force are such that the mass will just slide along at constant velocity, μ may be determined, where μ = effort/gravitational force. This simple experiment is somewhat limited in accuracy because the static friction is higher than limiting or dynamic friction.

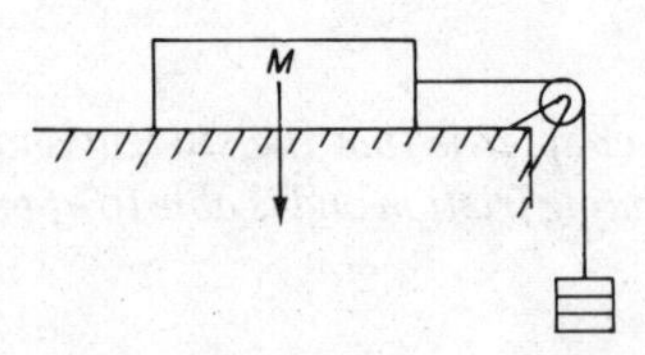

Figure 3.3 The relationship between a mass m being slid along a dry horizontal surface by a friction effort due to the gravitational attraction on known masses will enable the coefficient of friction to be determined

Strictly speaking, a particular coefficient of friction should not be applied to a single material but should always be applied to the two surfaces which are in contact.

Typical values of μ are

ice on ice	0.02–0.03
bronze on bronze (lubricated)	0.05
mild steel on mild steel	0.10
bronze on bronze (unlubricated)	0.20
cast iron on bronze	0.21
cast iron on brass	0.20
mild steel on bronze	0.34
hard steel on hard steel	0.42
mild steel on mild steel	0.60
wood on wood	0.25–0.50
wood on steel	0.50–0.55
leather on steel	0.55
leather on cast iron	0.56

Example 3.1

A crate of mass 500 kg is to be hauled along a horizontal floor by a force which is applied parallel to the floor (figure 3.4). (a) If the coefficient of friction between floor and crates is 0.6, determine the force necessary to move the crate. (b) If the force is then applied at an angle of 30° above the horizontal pulling the case along, determine the force necessary to produce movement.

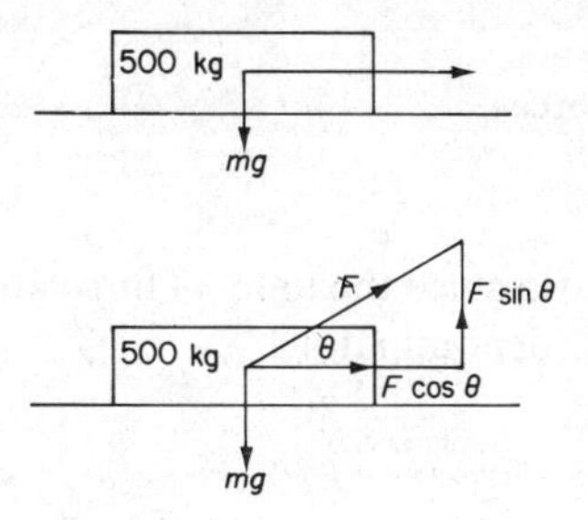

Figure 3.4

Solution (a)

$$\text{Friction force} = \mu mg$$
$$= 0.6\ (500 \times 9.81)$$
$$= 0.6 \times 4905$$
$$= 2943\ \text{N}$$

(b) The applied force must be split into two components: $F \cos \theta$ parallel to the surface and $F \sin \theta$ perpendicular to the surface. Only $F \cos \theta$ will tend to produce motion, therefore

$$F \cos \theta = \mu mg$$
$$= 0.6\ (500 \times 9.81)$$

$$F = \frac{0.6\ (500 \times 9.81)}{\cos 30} = \frac{0.6\ (500 \times 9.81)}{0.866}$$

$$= 3398\ \text{N}$$

The increased force is necessary because the force is not being applied efficiently.

The alert reader will realise that this is not the complete answer because we have neglected the vertical component ($F \sin \theta$) of the applied force. This will be tending to lift the crate off the floor thereby tending to reduce the friction resistance. Thus more correctly

$$\text{friction force} = \mu \times \text{total force normal to surface}$$
$$F \cos \theta = \mu(mg - F \sin \theta)$$
$$= \mu mg - F\mu \sin \theta$$

$$F \cos \theta + F\mu \sin \theta = \mu mg$$
$$F\ (\cos \theta + \mu \sin \theta) = \mu mg$$

$$F = \frac{\mu mg}{\cos \theta + \mu \sin \theta} = \frac{0.6\ (500 \times 9.81)}{0.866 + 0.3}$$

$$= \frac{2943}{1.166}$$

$$= 2524\ \text{N}$$

This confirms our everyday experience that it is easier to pull something along the floor rather than push it, since when pulling upwards we are tending to reduce friction whereas when pushing we tend to increase friction.

Example 3.2

The table of a machine tool weighs 4700 N and carries a component and fixture with a combined mass of 100 kg. If the coefficient of friction between the slides is 0.06, determine the total force necessary to move the table against the friction and a cutting force of 180 N.

Solution

$$\text{Total force on slideways} = 4700 + (100 \times 9.81)$$
$$= 4700 + 981$$
$$= 5681\ \text{N}$$

Frictional resistance $= \mu mg$
$= 0.06 \times 5681$
$= 341$ N

Total force necessary = friction force + cutting force
$= 341 + 180$
$= 521$ N

The importance of friction (or rubbing) and its associated problem of wear are so important in our technological age that a comparatively new science, that of *Tribology*, has been created to deal with problems created by man in his endeavour to control friction and reduce wear. Man has of course been attempting to minimise friction since the introduction of the wheel (from 3500 B.C.) and friction is vital to us in all aspects of our life: to walk we require friction (try walking on an ice-slide) and to play our games (spinning a ball through the air). Arthritis is a condition which results from the breakdown of lubrication in body joints.

The study of friction and wear spreads across several fields of science and since it has been estimated in the United Kingdom and in the United States that losses due to these causes are phenomenal (the Jost Report in 1965 estimated that the United Kingdom could save at least £515 m per annum (at 1965 values) by the application of better tribological practices), it behoves every technician engineer to recognise the importance of friction and the associated problems of wear, loss of efficiency and increased breakdowns.

We have already said that friction is not always a disadvantage but is often vital to our way of life. In engineering there are many applications where friction is vital to enable a mechanism to function. For example, the gripping of a component in a vice depends on increasing the normal reaction between the component and the vice jaws to increase the frictional grip. Power transmission devices such as belt and pulley drives and clutches depend for their operation on frictional grip between surfaces; in practice steps are taken to ensure good frictional adherence between appropriate surfaces, for example, correct belt tensions in belt drive mechanisms and the use of friction materials in clutch plates.

3.2 ANGLE OF FRICTION (ANGLE OF REPOSE)

We have established the relationship

$$\text{friction force} = \mu mg$$

thus the coefficient of friction, μ, between two surfaces can be expressed as

$$\mu = \frac{\text{friction force}}{mg}$$

(μ will be a pure ratio since the units of force are used in both the numerator and the denominator.)

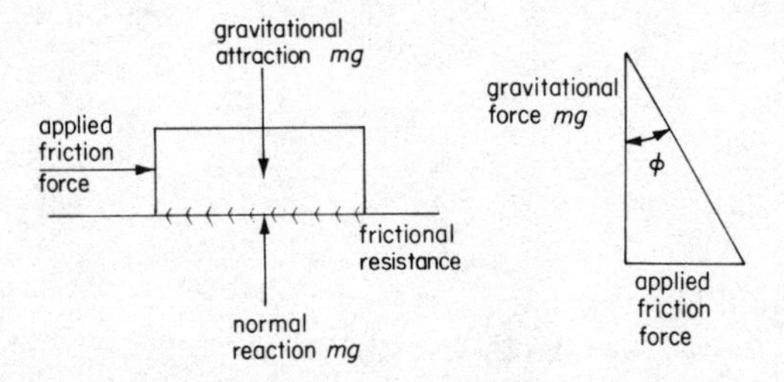

Figure 3.5 Applied and reaction forces acting on a mass sliding along a horizontal surface

Referring to figure 3.5, the ratio friction force/mg will also give the tangent of the angle ϕ, thus giving rise to an alternative method of expressing the coefficient of friction

$$\tan \phi = \mu = \frac{\text{friction force}}{mg}$$

This angular concept of μ gives rise to an alternative experimental method of determining the coefficient of friction between two surfaces by considering the effect of friction on an object resting on a plane inclined at an angle θ to the horizontal (figure 3.6). The gravitational attraction mg acting vertically downwards may be split or resolved into two components

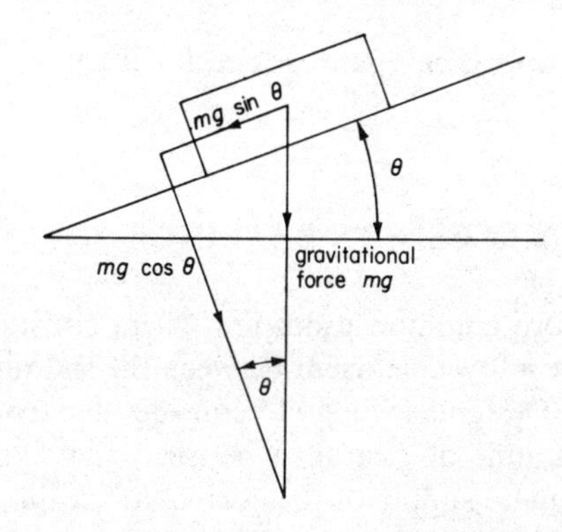

Figure 3.6 The gravitational attraction *mg* acting on a mass resting on an inclined plane may be split into two components: *mg* sin θ acting parallel to the plane and *mg* cos θ acting normal to the plane; *mg* sin θ will tend to produce motion down the plane against frictional resistance μ*mg* cos θ

(1) $mg \sin\theta$ acting down the plane tending to cause the object to slide down the plane against the frictional resistance, and

(2) $mg \cos\theta$ acting normal (at 90°) to the surface affecting the frictional resistance acting on the body by slightly reducing it; thus frictional resistance will equal $\mu mg \cos\theta$ rather than μmg.

The body will start to slide down the slope under the effects of its own gravitational attraction when the force tending to produce motion ($mg \sin\theta$) is equal to, or greater than, $\mu mg \cos\theta$ resisting motion. Thus

$$mg \sin\theta = \mu mg \cos\theta$$

$$\frac{mg \sin\theta}{mg \cos\theta} = \mu$$

$$\tan\theta = \mu$$

Thus if the angle of inclination of the surface is adjusted until constant-velocity motion just occurs, then the tangent of the angle θ at this instant will be equal to μ for those particular surfaces. This angle is known as the angle of repose and will, of course, be equal to the angle of friction ϕ. The accuracy of results so obtained will be limited since, once again, the effects of static friction will tend to prevent the body making its initial movement.

Example 3.3

Experimentally it is found that a block of mass 10 kg will just slide down an inclined plane with constant velocity when the angle of inclination is 27°. Find the force necessary to cause the block to slide over a horizontal surface of the same material (assuming the force is applied horizontally).

Solution

$$\begin{aligned}\mu &= \tan\text{ (angle of repose)}\\ &= \tan 27^\circ\\ &= 0.509\end{aligned}$$

$$\begin{aligned}\text{Friction force} &= \mu mg\\ &= 0.509 \times 10 \times 9.81\\ &= 49.93\text{ N}\end{aligned}$$

3.3 FULL-FILM AND BOUNDARY LUBRICATION

In those practical cases where friction must be reduced to a minimum, a lubricant may be introduced between the surfaces. The lubricant will ensure either partial or complete separation of the surfaces. When only partial separation of the surfaces is achieved (with a consequent reduction in frictional resistance), metal-to-metal contact will still exist at the 'peaks' (figure 3.7b) with the consequent danger of seizure still remaining (due to localised welding at these points of extreme pressure). This is often known as boundary lubrication.

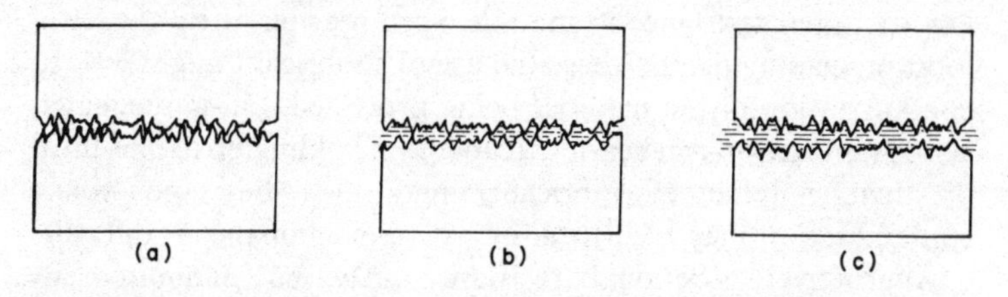

Figure 3.7 (a) Metal-to-metal contact between dry surfaces in contact; (b) boundary lubrication with partial separation between the surfaces; (c) full-film lubrication with complete separation of the surfaces

In more critical cases the lubricant may be pressure fed, and this, coupled with correct bearing design and an adequate supply of the correct lubricant, will ensure complete separation of the surfaces. In this case of full-film lubrication (figure 3.7c), the only resistance to motion will depend on the viscosity of the lubricant, which in turn depends on the resistance to shear of the lubricant, since the relative motion of the surfaces and the adherence of the lubricant to the surfaces will bring about shear in the lubricant film.

The viscosity of a lubricant may be affected by the operating temperature and a lubricant must be selected (or blended) to give the correct viscosity at the prevailing temperatures. The lubricant must also have an adequate film strength to support the load without breaking down thereby permitting metal-to-metal contact. The oiliness of the lubricant, that is, the ease with which lubricated surfaces slide over each other, is of prime importance in the case of boundary lubrication. Other important properties of a lubricant include heat-dissipation characteristics (in oils), resistance to foaming (greases), chemical stability and availability at reasonable cost.

Lubricants may be in liquid form (mineral oils, fixed or natural oils, soluble oils and synthetic fluids), in plastic form (greases), solid (graphite, molybdenum disulphide) or gaseous (air).

Frictional resistance between surfaces may also be minimised by careful choice of materials. The ideal bearing material will consist of a soft matrix or base material in which harder load-carrying particles are embedded. The use of inserts, or bearing pads, enables control to be exercised over the nature of mating surfaces.

The use of plastics materials, such as nylon and PTFE, will enable bearings to be run dry (without lubrication) and still give low frictional resistance to motion. Such bearings are an asset in food-processing plants where the use of a lubricant might lead to contamination of the material being processed. Oil-impregnated porous bronze or iron bearings (capable of holding up to one-third of their volume of lubricant) may also be used where contamination may be critical or where maintenance is difficult.

Alternatively frictional resistance may be minimised by redesigning to replace sliding motion between mating surfaces with rolling motion by the use of rolling-bearings. Lubrication will still be required, but where contamination is critical the bearings may be lubricated (with grease) and sealed for life.

3.4 FRICTION LOSSES IN BEARINGS

One of the more common examples where considerable care is taken to ensure a low coefficient between the sliding surfaces is in shaft bearings. Designers will go to considerable trouble to obtain the correct amount of clearance between shaft and bearing to ensure an adequate supply of the correct lubricant, at the correct pressure, to ensure complete separation of the sliding surfaces and minimise losses due to friction.

A stationary shaft lying in its bearing makes metal-to-metal contact at the bottom of the bearing, as shown in figure 3.8a. As clockwise rotation begins, the friction effect of the metal-to-metal contact will cause the shaft to climb further to the right (figure 3.8b) and a wedge of oil will be drawn between the shaft and the bearing. An increase in the shaft speed causes it to slip in the bearing and the increasing pressure being built up in the wedge of lubricant displaces the shaft to the left. If the bearing is hydrodynamically correct, no metal-to-metal contact occurs, full-film lubrication is achieved and the frictional resisting torque will be due to the viscosity of the lubricant.

In less critical cases the designer will call for the provision of oil holes and grooves to ensure the distribution of the lubricant throughout the bearing, when conditions of boundary lubrication will be achieved.

The torque or power lost in overcoming the frictional resistance to motion in shaft bearings can be determined by applying horizontal surface friction theory to the angular or rotational situation. Assuming that the shaft is resting at the bottom of its bearing while it is rotating, then, by analogy, the friction force acting between the shaft and the bearing will be (figure 3.9)

$$\text{friction force} = \mu mg$$

This will give a frictional torque of μmgr N m, where r is the radius of the shaft (m).

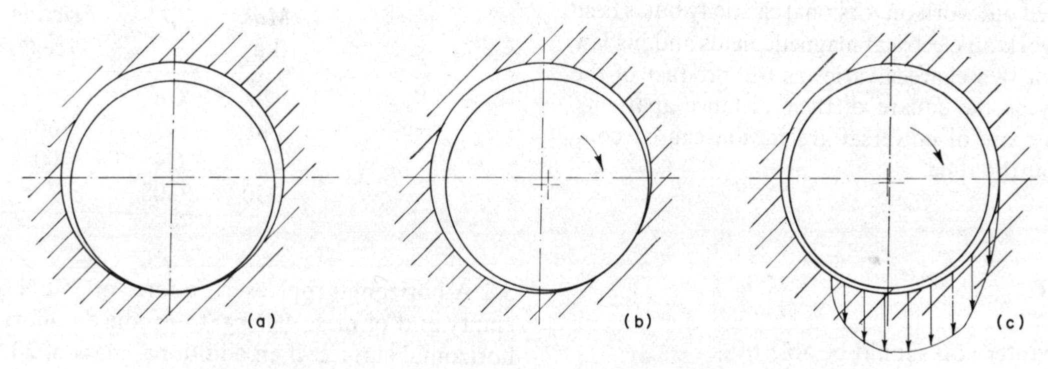

Figure 3.8 (a) Stationary shaft lying in its bearings; (b) clockwise rotation of shaft causes shaft to climb to the right; (c) increasing pressure in the wedge of lubricant causes the shaft to be displaced to the left

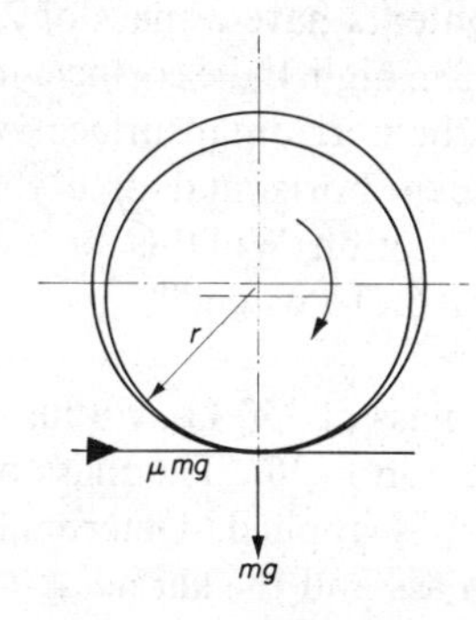

Figure 3.9 Frictional torque in a bearing is given by the frictional resistance *mg* multiplied by the radius of the shaft

Example 3.4

A shaft and rotor with a combined mass of 2000 kg rotates in a bearing of 50 mm diameter. If the coefficient of friction between shaft and bearing is 0.08, determine the torque necessary to overcome the friction resistance.
Solution

$$\begin{aligned}\text{Frictional torque} &= \mu mgr \\ &= 0.08 \times 2000 \times 9.81 \times 0.025 \\ &= 39.2 \text{ N m}\end{aligned}$$

(This will be the same no matter how many bearings are used, because the total load carried in all the bearings will remain the same.)

Example 3.5

A drum 1 m in diameter has a brake block lined with a material to give a coefficient of friction between brake and drum of 0.6. If a force of 700 N is applied to the drum, find the braking torque.
Solution

$$\begin{aligned}\text{Braking torque} &= \mu mgr \\ &= 0.6 \times 700 \times 0.5 \\ &= 210 \text{ N m}\end{aligned}$$

1 Charles Coulomb, whose name is commemorated in the name given to the unit of electrical charge, lived in Paris from 1736 to 1806. In 1781 he verified work carried out earlier by another Frenchman, G. Amontons, in distinguishing between static and dynamic sliding friction, and suggested that frictional resistance might be due to molecular adhesion, a theory which has been shown to be very largely true.

Coulomb also carried out work on torsional elasticity but is best remembered for his work on electrical magnetic fields and his law that the force between two charges varies as the product of the charges and inversely as the square of their distance apart; its similarity to Newton's law of universal gravitation caused considerable excitement at the time.

TO THE STUDENT

At the end of this chapter you should be able to

(1) explain the nature of the frictional resistance between two sliding surfaces
(2) appreciate the factors which affect the frictional resistance
(3) calculate the force necessary to overcome friction using the formula $F = \mu mg$
(4) calculate the torque necessary to overcome frictional resistance using the formula $T = \mu mgr$
(5) complete exercises 3.1 to 3.13.

EXERCISES

3.1 Explain what is meant by friction. Give three examples whereby friction is used advantageously in engineering.

3.2 Explain the causes of the resistance to dry sliding motion between two mating surfaces. How may this resistance be varied?

3.3 In some practical cases it is desirable to minimise friction between two sliding surfaces. Give three examples of these instances stating how the friction may be minimised, and the effect if, under these circumstances, friction were not minimised.

3.4 Complete the following table, assuming that the mass is being moved over a horizontal surface by a force which is applied horizontally.

Mass (kg)	μ	*Friction Force* (N)
20	0.6	
40		160
	0.4	200
250	0.08	

3.5 A horizontal rope exerts a force of 350 N on a crate, initially empty and of mass 400 kg, to produce uniform motion across a horizontal surface. If an additional mass of 20 kg is added to the crate, what then will be the force exerted by the rope to produce motion?

3.6 A crate and its contents have a mass of 250 kg, being 0.4 m wide, 0.8 m long and 0.3 m high. If the coefficient of friction between the wooden crate and the horizontal surface over which it is to be moved is 0.2, determine the horizontal force which must be applied to the crate to produce motion. Will this force be altered if the case is turned on its side or on to its end?

3.7 It is found that a mass of 250 kg will just slide down a plane whose angle of inclination is 36°. The mass moves with uniform velocity and no effort is applied. Determine the coefficient of friction between the mass and the surface.

3.8 A mass of 90 kg will just slide down an incline of 20° with uniform velocity and without the application of any force. What effort would be required to prevent the same block sliding down an incline of 36°?

3.9 A mass of 150 kg is pulled along a horizontal surface by means of an effort applied at 20° to the surface. If the coefficient of friction is 0.12, determine the effort required to produce constant velocity motion.

3.10 A labourer pushing downwards at 30° to the horizontal with a force of 400 N is able to move a mass of 25 kg along a horizontal surface with a constant velocity. Determine the coefficient of friction between the mass and the surface.

3.11 A shaft assembly whose mass is 200 kg is supported in bearings of 30 mm diameter. If μ between the shaft and the bearing is 0.05, determine the turning moment required to overcome friction.

3.12 Calculate the frictional torque in a bearing 80 mm diameter if it is supporting a load of 2000 kg and $\mu = 0.05$.

3.13 It is found that the frictional torque in a bearing is 0.035 N m. If the bearing is 40 mm diameter and μ is 0.08, determine the load supported by the bearing.

NUMERICAL SOLUTIONS

3.4 117.7 N; 0.4; 51 kg; 196 N
3.5 367 N
3.6 490.5 N; no
3.7 0.726
3.8 260 N
3.9 188 N
3.10 0.778
3.11 1.47 N m
3.12 39.24 N m
3.13 22.4 kg

4 Simple Machines

The object of this chapter is to introduce the student to simple machines and to make him aware of the factors affecting the efficiency of such machines.

4.1 SIMPLE MACHINES AND LEVERS

A machine may be defined as a device or means for converting energy from one form to another; for example, an electric motor converts electric energy into mechanical energy, a dynamo converts mechanical energy into electrical energy, a petrol engine converts the chemical energy of the fuel into heat and mechanical energies, and a conventional machine tool receives electrical energy into the motor and produces mechanical energy at the tool point.

Since energy is the ability to do work, and work done and energy both have the same units (joules), a machine can also be said to be a device or means of converting work done from one form to another.

Until recently, engineers and the general public alike have taken for granted the availability of cheap and plentiful supplies of energy, but recent political and economic events have made our base sources of energy—fossil fuels like oil and coal—more costly and difficult to obtain. Even with the advent of supplies of oil from the North Sea, the cost of extraction of this oil and the current world prices will ensure that the cost of energy will remain comparatively high. Coal supplies in the United Kingdom, being sufficient for our present needs, are likely to prove more expensive to extract; coupled with this are the more important costs of extraction in terms of damage to the countryside. Thus we must consider the environmental and conservation issues involved in the extraction, transport, and use of fuels and energy. It is important to ensure that energy-conversion processes are carried out with as little loss as possible and this means that all machines must be operated as efficiently as possible.

At this early point in our studies we are only concerned with simple machines like elementary lifting devices. These are usually machines which enable heavy loads to be lifted by the application of much smaller effort. Both the input energy and the output energy are in the form of mechanical energy and therefore both input and output from the machine can be expressed as a force being moved through a distance. When considering the input, the force will be the effort being put into the machine; when considering the load, the force will be the load or the resistance to the machine.

Probably the most simple example of all lifting machines is the lever. All of us at some time or other have used a lever as a means of moving a load that would otherwise be beyond our capabilities—a labourer may use a crowbar to move a heavy machine, and a thief may use the same implement for other purposes. A wheelbarrow enables us to lift a large mass in the barrow by applying a smaller effort at the handles.

However, using the lever is not entirely advantageous gain: we find that we have to move our smaller effort through a much greater distance than the distance through which the heavy load is moved (figure 4.1).

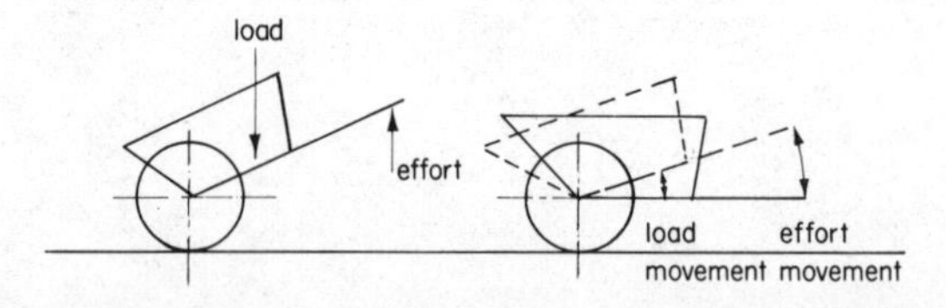

Figure 4.1 A wheelbarrow is an example of a simple lever or machine enabling a heavy load to be raised and moved by means of a smaller effort, but the distance through which the effort must be moved is greater than the distance through which the load is moved

There are, however, situations in which levers are used when the load is not greater than the effort. On a symmetrical (evenly balanced) see-saw, for instance, the load being lifted would be the same as the downward acting effort, in which case the distances moved through would be identical. Here the simple machine would only be used to achieve a change in direction of the movement.

Again there are cases, albeit fewer, where a simple lever may be used to multiply the distance being moved. Thus a large effort may be applied and moved through a small distance to move a smaller load through a much larger distance. Examples of this occur in the human body, for example, where the biceps is used to lift the lower arm (and whatever load it may be carrying).

We may classify levers according to application, as outlined in the three conditions above, but when referring to machines generally no such classification is used.

4.2 MECHANICAL ADVANTAGE AND VELOCITY OR DISTANCE RATIO

We have seen that we use a lifting machine to give the advantage of being able to lift a heavy load while applying a small effort. All machines are used to gain some advantage and it is useful when comparing machines to be able to express such an advantage in numerical terms. This may be done by using the term *mechanical advantage* where

$$\text{mechanical advantage} = \frac{\text{load or resistance}}{\text{effort}}$$

Mechanical advantage is a ratio (and therefore has no units) and so it is important that both the load (or resistance) and the effort be measured in the same units. This is particularly important when determining the mechanical advantage of lifting machines, where the load may be specified in terms of mass (kilograms) and the effort in terms of force (newtons); conversion must be carried out in such a case.

The numerical value of the mechanical advantage ratio is usually greater than unity for most machines. It most certainly is greater than 1 for any lifting machine, where the load will always be greater than the effort. Furthermore, this ratio will vary for any particular machine depending on the conditions under which the machine is being used. This can easily be appreciated if we realise that the input effort to a machine must be sufficient to do two things: (1) move the load or overcome the resistance (as in a cutting tool) and (2) overcome the inertia and resistance of the machine. Obviously, if a machine is not lubricated or maintained effectively, then a greater proportion of the input effort must be devoted to overcoming the considerable frictional resistance of the machine. Consequently a smaller proportion of the effort will be available to move the load, which must therefore be smaller, giving a reduced mechanical advantage.

Similarly, when a machine is used under conditions for which it has not been designed, the mechanical advantage will be reduced. For instance, if a crane, which is designed to be strong and massive to lift loads of 5 tonnes safely, is used to move smaller loads, say, 500 kg, then a large proportion of the input effort must be devoted to overcoming the considerable inertia of the heavy machine, giving a poor mechanical advantage. If the machine were used under the conditions for which it was designed—lifting heavier loads—although the frictional resistance of the machine will increase slightly, the inertial resistance of the machine, due to the massiveness of the machine, will be the same and consequently a greater proportion of the increased effort will be available to move the load, thus giving an increased mechanical advantage.

From this we can see that the mechanical advantage is an indication of the benefit gained from using a machine, and is not constant for a particular machine. This will be shown graphically in figure 4.11.

However, we have also seen that we must move our effort through a greater distance than that through which the load or resistance moves. The ratio of these distances is decided by the design of the machine and remains a constant for that particular machine no matter under what conditions the machine is used. This ratio, correctly known as the distance ratio, is given by

$$\text{distance ratio} = \frac{\text{distance moved by effort}}{\text{distance moved by load}}$$

The same ratio has, in the past, been known as the velocity ratio, which is acceptable since in any given time interval the distances moved will always be proportional to the velocities involved. The distances will be fixed by the design of the machine and therefore the distance or velocity ratio will be a constant for any given machine no matter what the conditions of use.

Example 4.1

A machine raises a load of 82 kg through a distance of 250 mm by an effort of 110 N which is moved through a distance of 2.25 m. Determine the mechanical advantage at this loading and the velocity ratio of the machine.

Solution

$$\text{Mechanical advantage} = \frac{\text{load}}{\text{effort}}$$

$$= \frac{82 \text{ kg}}{110 \text{ N}}$$

$$= \frac{82 \times 9.81}{110}$$

$$= 7.3$$

$$\text{Velocity ratio} = \frac{\text{distance moved by effort}}{\text{distance moved by load}} = \frac{2250}{250} = 9$$

4.3 VELOCITY RATIO OR DISTANCE RATIO OF SIMPLE MACHINES

We have said that the velocity ratio or distance ratio for a machine is constant and will depend on the design and method of operation of the machine. By considering the design details of simple machines it is possible to determine standard formulae to calculate the ratios.

4.3.1 Wheel and Axle

(See figure 4.2.) This device involves a small effort being applied at

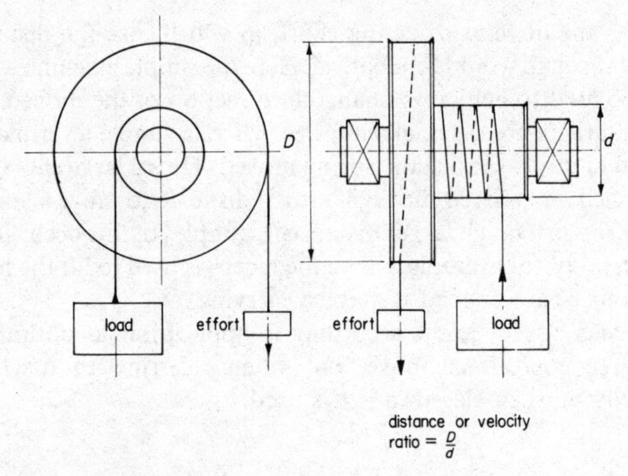

Figure 4.2 A simple wheel and axle

the end of a long lever or handle (or a larger diameter pulley) while the load is supported on the rope wrapped around the axle; the whole assembly is supported on bearings.

Considering one revolution of the assembly, the effort will travel through a distance equal to the circumference of the large pulley, πD, (or the circumference of a circle whose radius is the length of the lever or handle) and the load will be lifted up a distance equal to the circumference of the axle (ignoring the thickness of the cord) πd. Thus

$$\text{distance or velocity ratio} = \frac{\text{distance moved by effort}}{\text{distance moved by load}}$$

$$= \frac{\pi D}{\pi d}$$

$$= \frac{D}{d}$$

that is, the ratio between the diameters.

4.3.2 Differential Wheel and Axle

(See figure 4.3.) This is an extension of the simple wheel and axle,

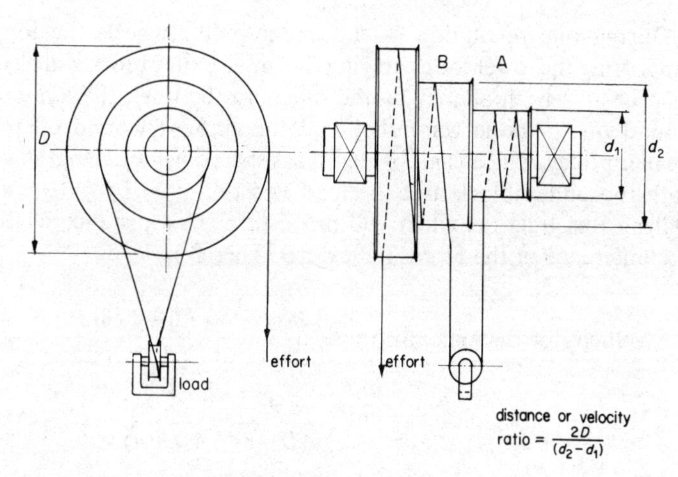

Figure 4.3 Differential wheel and axle

where three different diameters are used, with the cord supporting the load being wrapped around the two smaller diameters. However, the directions in which the cord is wound are reversed on the two diameters, so that as the axle assembly rotates the rope is unwound on one axle and wound on to the other axle; the load is supported from a pulley assembly between the two 'falls' of rope.

Considering one revolution during which the load is being raised

length of rope unwound from axle A $=\pi d_1$
length of rope wound on to axle B $=\pi d_2$
falls of the rope decrease by $\pi(d_2-d_1)$

This decrease in length must be shared equally between the two falls of rope thus the load will be raised during one revolution by

$$\frac{\pi(d_2-d_1)}{2}$$

During this same revolution the effort will be moved through

$$\pi D$$

Thus

$$\text{distance or velocity ratio} = \frac{\text{distance moved by effort}}{\text{distance moved by load}} = \frac{\pi D}{\frac{1}{2}\pi(d_2-d_1)} = \frac{2D}{(d_2-d_1)}$$

Thus the differential wheel and axle can be used to give a much higher distance or velocity ratio than the simple wheel and axle.

4.3.3 Pulley Blocks

(See figure 4.4.) These are among the most common of all lifting devices and consist of a single rope or chain, which passes around a number of pulleys or sheaves mounted in pulley blocks.

The distance through which the effort is moved will shorten the over-all length of rope running over the pulleys, causing the load to

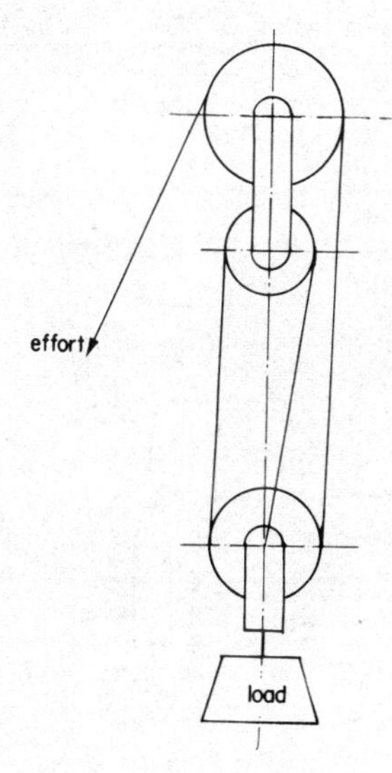

Figure 4.4 Simple pulley block system

be raised by a smaller amount depending on the number of falls of rope involved. The number of falls of rope will be the same as the number of pulleys involved. For example, with three pulleys involved and the effort being moved through 0.3 m, the three falls of rope will be shortened by 0.3 m hence the load will be raised through 0.1 m. Thus the velocity ratio will be given by

$$\text{velocity ratio} = \frac{\text{distance moved by effort}}{\text{distance moved by load}}$$

$$= \frac{0.3}{0.3/3} = 3$$

$$= \text{no. of falls of rope} = \text{no. of pulleys}$$

4.3.4 Weston Differential Pulley Block

(See figure 4.5.) This uses an endless chain or rope wound over an upper pulley block, consisting of two pulleys of differing diameters, and a lower pulley block carrying the load, with a loop as shown, where the effort is applied.

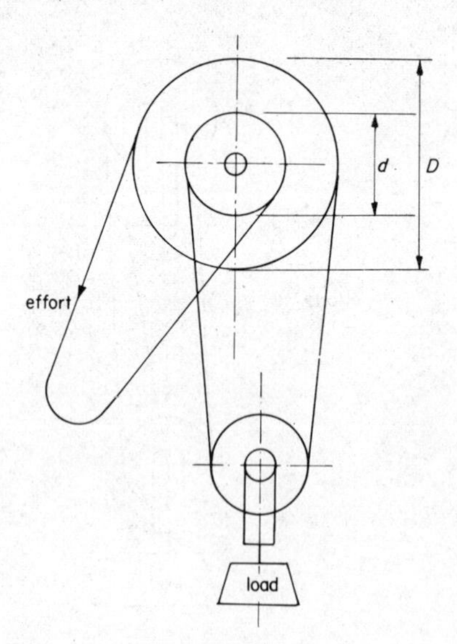

Figure 4.5 Weston differential pulley block

During one revolution of the upper pulley wheels the loop supporting the lower loop carrying the lower pulley block will have its over-all length altered by the difference between the amount wound on to the larger pulley and the amount wound off the smaller pulley, which is $\pi(D-d)$. This is shared by both lengths, so both the pulley block and the load will be raised by $\frac{1}{2}\pi(D-d)$. During this time the effort will move through a distance of the circumference of the larger pulley, πD. Therefore

$$\text{velocity or distance ratio} = \frac{\text{distance moved by effort}}{\text{distance moved by load}}$$

$$= \frac{\pi D}{\frac{1}{2}\pi(D-d)} = \frac{2D}{(D-d)}$$

This mechanism again will give a higher velocity ratio than a simple pulley block arrangement and can be maximised if the diameters of the upper pulleys are almost equal.

Sometimes, for heavier loads, chains may be used instead of ropes, and in such cases the pulleys must be designed to accommodate the links of the chain. This is done by providing the pulley with teeth or flats to engage with the chain, the number of teeth or flats being proportional to the pulley diameter; the expression for the velocity ratio can then be expressed in terms of number of flats or teeth thus

$$\text{velocity ratio} = \frac{2N}{(N-n)}$$

4.3.5 The Screw Jack

(See figure 4.6.) This consists of a table which may be raised or lowered by a screw thread turning in a fixed body or frame. The screw is rotated by an effort applied at the end of a handle or tommy bar. Alternatively the effort may be applied by means of a cord wrapped around the outside of the circular table (this method is often used in laboratory equipment).

The rate at which the load carrying the table or platform is lifted will depend on the pitch of the screw thread and the number of

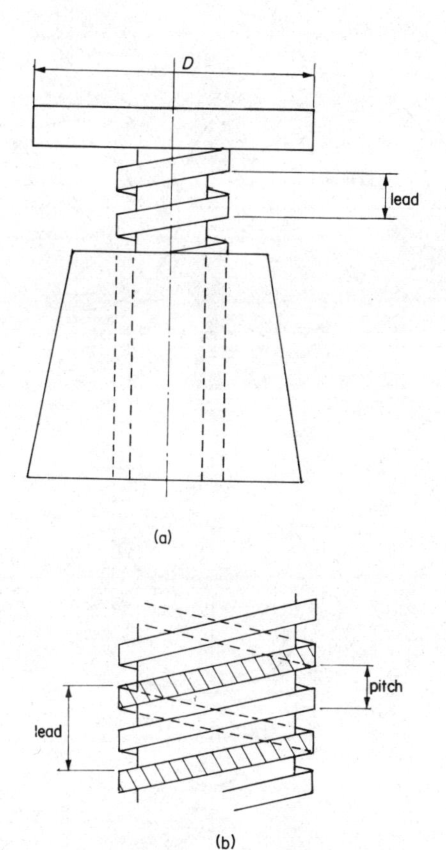

Figure 4.6 (a) Screw jack with single-start thread; (b) multi-start thread, where lead = $n \times$ pitch (n = number of starts)

'starts'. Consider a 'simple' screw thread with a single-start thread (figure 4.6a), then in one revolution the screw will move axially a distance equal to the pitch (lead = $1 \times$ pitch). However, in a double-start screw, which really is two parallel threads (figure 4.6b) the load or axial movement is twice the pitch; thus for a multi-start thread the lead = $n \times$ pitch, where n = number of starts.

This means that, considering one revolution of the screw

$$\text{velocity or distance ratio} = \frac{\text{distance moved by effort}}{\text{distance moved by load}}$$

$$= \frac{\pi D}{\text{lead}} = \frac{2\pi R}{\text{lead}}$$

where D = diameter of the table and, R = radius of the handle.

4.3.6 Worm and Worm Wheel

(See figure 4.7.) Here the load is lifted by a pulley on the same axle as the worm wheel. This worm wheel meshes with the worm, which may be regarded as a screw, which is driven by the effort wheel.

If the worm has a single-start 'thread' and the wheel has N teeth, it will require N revolutions of the worm to produce 1 revolution of the wheel. If the worm is a multi-start worm, say, n starts, then the worm wheel will rotate more quickly (since the 'speed' of the thread will be increased) and each revolution of the worm wheel will require N/n revs of the worm. The velocity ratio of the worm and

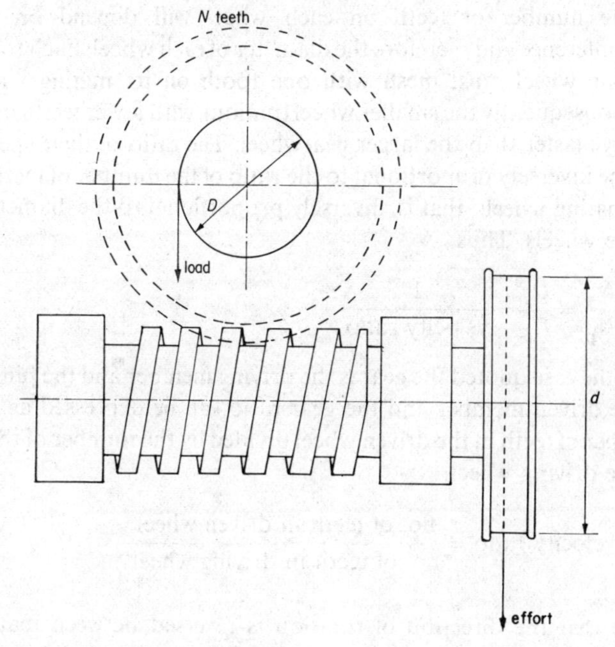

Figure 4.7 Worm and worm wheel

worm wheel will therefore be N/n and if D is the diameter of the load pulley and d the diameter of the effort wheel and pulley the velocity ratio of the device will be

$$\text{velocity ratio} = \frac{N(\pi d)}{n(\pi D)} = \frac{Nd}{nD}$$

4.3.7 Simple Gear Train

(See figure 4.8.) A simple gear train is used, either as part of a more complex machine or as a device in its own right, to transmit power positively between two or more shafts lying parallel to each other.

The smaller of the two gears in mesh is referred to as the pinion and the term spur or gearwheel is applied to the larger wheel. The pinion and gearwheel have similar sized projections or teeth arranged on their periphery to ensure positive transmission of the power; these teeth are shaped to ensure that the teeth roll together (rather than slide over each other) to give more efficient transfer of power.

The number of teeth on each wheel will depend on the circumference and therefore the diameter of each wheel. Each tooth on one wheel must mesh with one tooth on its mating wheel and consequently the smaller wheel (pinion), with fewer teeth, must revolve faster than the larger gearwheel. The ratio of their speeds will be inversely proportional to the ratio of the number of teeth in the mating wheels, that is, inversely proportional to the diameters of the wheels. Thus

$$\frac{N_g}{N_p} = \frac{D_g}{D_p} = \frac{1}{\text{velocity ratio}}$$

In the case quoted the gear is the driving member and the pinion is the driven member and the gear ratio can be expressed as the number of teeth in the driven wheel divided by the number of teeth in the driving wheel

$$\text{velocity ratio} = \frac{\text{no. of teeth in driven wheel}}{\text{no. of teeth in driving wheel}}$$

Note that the direction of rotation is reversed between mating gearwheels.

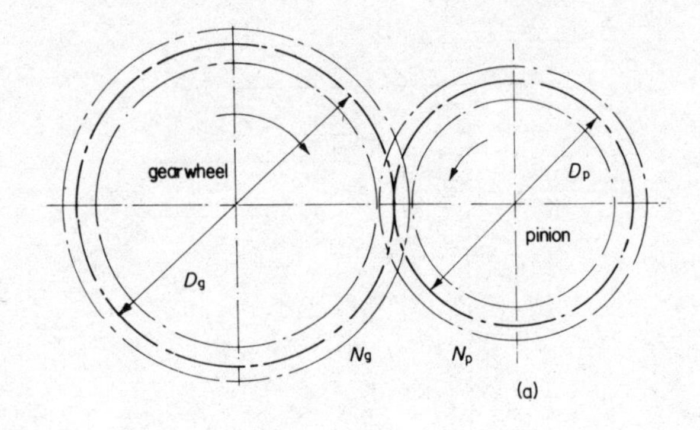

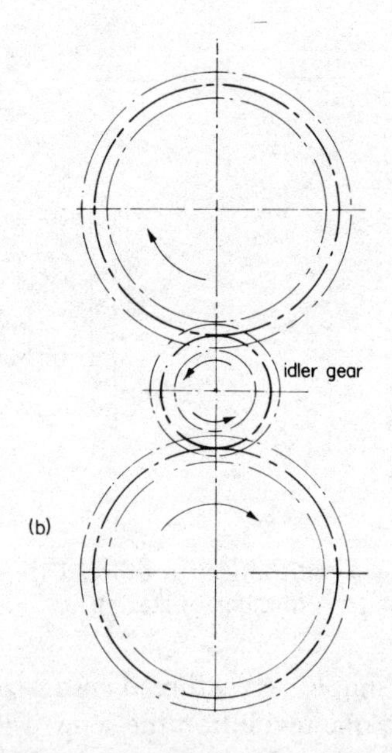

Figure 4.8 (a) Simple gear train; (b) simple gear train with idler gear to restore original direction of rotation without causing any change in gear ratio

The introduction of a third wheel (figure 4.8b) between two mating gears will have the effect of reversing the direction of rotation to ensure that the output direction will be the same as the input. This third wheel will not alter the total ratio of the gear train since its effect is self-cancelling. This gives rise to the term 'idler' gear.

4.3.8 Compound Gear Train

(See figure 4.9.) In many practical cases where it is required to transmit power between two closely adjacent shafts there may be insufficient room to accomplish the required gear ratio using a simple gear train. The use of a 'compound' gear train, where two gearwheels are mounted on the same axle, can be used to achieve a larger gear ratio within a small centre distance.

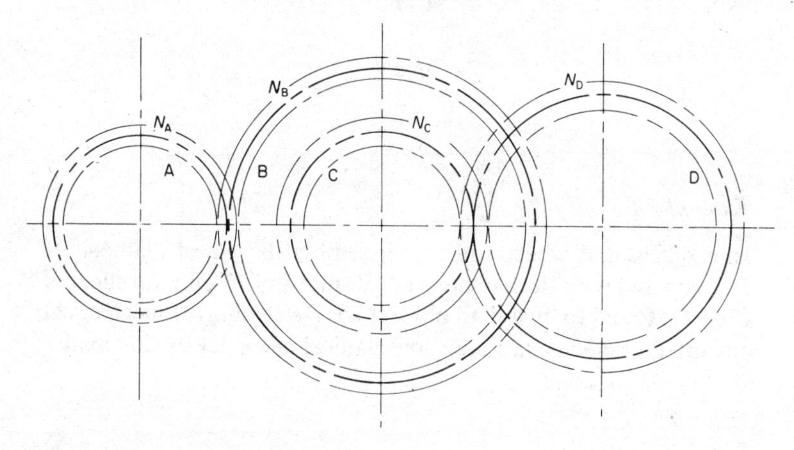

Figure 4.9 Compound gear train

Considering figure 4.9, the driving wheel A meshes with B giving a ratio of N_A/N_B; thus for 1 revolution of A, B makes only a part of a revolution. The pinion C is fixed to the same axle, so it will make the same number of revolutions; hence $(N_A/N_B) \times N_C$ teeth will pass the point at which pinion C meshes with gear D; thus D will make $(N_A/N_B) \times (N_C/N_D)$ revolutions, thus

$$\text{velocity ratio} = \frac{N_B}{N_A} \times \frac{N_D}{N_C}$$

Obviously the gear ratio of a compound gear train will be given by the product of the individual velocity ratios and pinions, which in each case is the number of teeth in the driven wheel divided by the number of teeth in the driver.

4.4 MECHANICAL EFFICIENCY OF SIMPLE MACHINES

Since we have defined a machine as a device for converting energy or work done from one form to another, it follows that the input to the machine and the output from the machine must both be some form of energy or work. Furthermore, since we know that energy cannot be created or destroyed, it follows that the output energy or work from the machine must always be less than the input, since the machine is not capable of creating energy or work. The machine does not destroy energy either, but some of the energy will be consumed or used up in driving the machine, overcoming the frictional and inertial resistances (section 4.2).

If we can arrange for the proportion of the input energy that is devoted to overcoming the resistances of the machine to be as small as possible, so that the output is as near in value to the input as possible, then the machine is said to be running with the highest possible efficiency, which we have already said is important for the sake of energy conservation.

Steps that we can take to improve or maintain the efficiency of machines include ensuring that an efficient programme of maintenance is carried out, adequate and suitable lubrication is provided (see section 3.4) and machines are used properly and safely for the correct purpose (see section 4.2).

We can measure the efficiency of a machine by comparing the output to the input, where

$$\text{mechanical efficiency} = \frac{\text{output}}{\text{input}}$$

Since the output must be less than the input, this means that the efficiency will always be less than unity. To save the inconvenience of using decimals it is more usual to refer to the percentage mechanical efficiency, obtained by multiplying the mechanical efficiency by 100. Thus

$$\text{mechanical efficiency } (\%) = \frac{\text{load or resistance} \times \text{distance moved by load or resistance}}{\text{effort} \times \text{distance moved by effort}} \times 100$$

We have already said that mechanical advantage = load/effort and velocity ratio = distance moved by effort/distance moved by load, therefore

$$\text{mechanical efficiency } (\%) = \text{mechanical advantage} \times \frac{1}{\text{velocity ratio}} \times 100$$

$$= \frac{\text{mechanical advantage}}{\text{velocity ratio}} \times 100\%$$

We have previously said that the mechanical advantage will vary, depending on the loading conditions imposed on any particular machine, while the velocity ratio remains a constant for any particular machine; it follows from the last expression that the mechanical efficiency must also vary according to the loading conditions.

Example 4.2

A wheel and axle has a wheel diameter of 250 mm and an axle diameter of 60 mm. A load of 55 kg is lifted by an effort of 230 N. Determine the velocity ratio, the mechanical advantage and mechanical efficiency at this loading.

Solution

$$\text{Mechanical advantage} = \frac{\text{load}}{\text{effort}}$$

$$= \frac{55 \times 9.81}{230} = 2.35$$

$$\text{Velocity ratio} = \frac{\text{distance moved by effort}}{\text{distance moved by load}}$$

$$= \frac{D}{d} = \frac{250}{60} = 4.17$$

$$\text{Mechanical efficiency} = \frac{\text{mechanical advantage}}{\text{velocity ratio}}$$

$$= \frac{2.35}{4.17} = 0.564$$

$$= 56.4\%$$

Example 4.3

In a differential wheel and axle assembly the wheel diameter is 600 mm and the axle diameters are 50 mm and 76 mm. An effort of 670 N is found to lift a load of 1100 kg. Determine the mechanical advantage, velocity ratio and mechanical efficiency at this load.

Solution

$$\text{Velocity ratio} = \frac{\text{distance moved by effort}}{\text{distance moved by load}}$$

$$= \frac{2D}{(D-d)}$$

$$= \frac{2 \times 600}{(76-50)} = \frac{1200}{26} = 46.15$$

$$\text{Mechanical advantage} = \frac{\text{load}}{\text{effort}} = \frac{1100 \text{ kg}}{670 \text{ N}} = \frac{1100 \times 9.81}{670}$$

$$= 16.1$$

$$\text{Mechanical efficiency} = \frac{\text{mechanical advantage}}{\text{velocity ratio}}$$

$$= \frac{16.1}{46.15} = 0.3489 = 34.89\,\%$$

Example 4.4

A winch is constructed with an operating handle 300 mm long, driving a shaft with a 25 mm pinion, meshing with a 100-tooth gearwheel, driving a load drum of 120 mm diameter. If, when a load of 300 kg is being lifted, the efficiency is 60 per cent, determine the effort which must be applied to the handle.

Solution During 1 revolution of the handle, the load drum makes 25/100 = 0.25 revolutions, therefore

$$\begin{aligned}\text{distance moved by load} &= \pi d \times 0.25\\ &= \pi \times 120 \times 0.25\\ &= 94.25\ \text{mm}\end{aligned}$$

$$\begin{aligned}\text{Distance moved by effort in 1 revolution} &= 2\pi \times \text{length of handle}\\ &= 2 \times 300\\ &= 1885\ \text{mm}\end{aligned}$$

therefore

$$\text{Velocity ratio of winch} = \frac{\text{distance moved by effort}}{\text{distance moved by load}}$$

$$= \frac{1885}{94.25}$$

$$= 20$$

$$\text{Mechanical efficiency} = \frac{\text{mechanical advantage}}{\text{velocity ratio}}$$

$$\begin{aligned}\text{Mechanical advantage} &= \text{mechanical efficiency} \times \text{velocity ratio}\\ &= 0.6 \times 20\\ &= 12\end{aligned}$$

Now mechanical advantage = load/effort, therefore

$$\text{effort} = \frac{\text{load}}{\text{mechanical advantage}} = \frac{300 \times 9.81}{12}$$

$$= 245.25\ \text{N}$$

4.5 LIMITING EFFICIENCY—THE LAW OF A SIMPLE MACHINE

The variation in performance of a simple machine is best illustrated by a series of graphs based on information obtained by carrying out a series of tests on a machine using various loads. The efforts required to move the loads are recorded and may be plotted on a graph, as shown in figure 4.10. The load is plotted horizontally and the effort vertically; the effort is shown to vary directly with the load, producing a straight line. This linear relationship between the

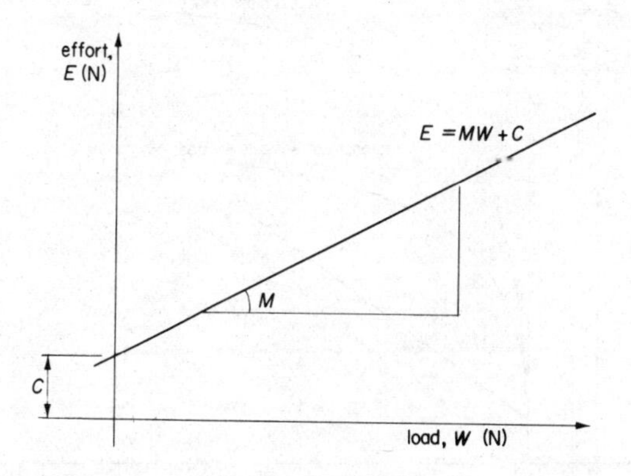

Figure 4.10 The linear relationship between load and effort enables the law of the machine to be found: $E = MW+C$

variables may be represented by

$$E = M\,W + C$$

which is the standard equation for any straight line where M is the slope (steepness) of the line and C is the intercept on the vertical axis, when $E = 0$. The actual values of M and C can be determined from the graph as illustrated (or by solving simultaneous equations when two corresponding loads and efforts are known). The intercept on the vertical axis when the load = 0 (the value of C) is the effort required to drive the machine when the load is zero, that is, the effort required to drive the unloaded machine, to overcome the friction and inertia.

This equation relating the effort and load is known as the *law of the machine.*

For small loads, the proportion of the effort taken up in overcoming friction must result in the machine operating at a lower efficiency. As the load increases the frictional resistance also increases, but the proportion of the input effort needed to overcome this resistance is lower, thus the efficiency is increased. (This is shown on figure 4.11). However, this increase in efficiency does not continue indefinitely as the load increases towards the maximum design load. The rate at which the efficiency increases

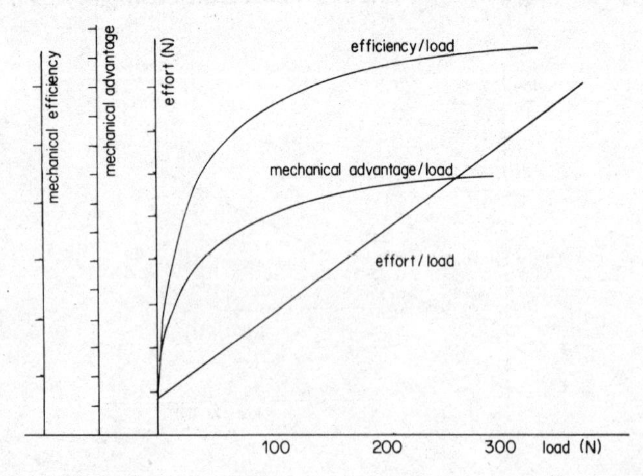

Figure 4.11 Graph showing variation of mechanical advantage and mechanical efficiency with load

becomes smaller and the efficiency reaches a limiting figure, which is known as the *limiting efficiency of the machine.* This limiting efficiency may be expressed mathematically as

$$\text{efficiency} = \frac{\text{mechanical advantage}}{\text{velocity ratio}}$$

$$= \frac{\text{load}}{\text{effort}} \times \frac{1}{\text{velocity ratio}}$$

but effort $= M\,W + C$, therefore

$$\text{efficiency} = \frac{W}{(M\,W + C)\ \text{velocity ratio}}$$

$$= \frac{W}{M\,W(\text{velocity ratio}) + C(\text{velocity ratio})}$$

dividing numerator and denominator by W

$$\text{efficiency} = \frac{1}{M \times \text{velocity ratio} + C \times \text{velocity ratio}/W}$$

As the load increases, the term $C \times$ velocity ratio/W gets smaller, so the denominator gradually approaches $M \times$ velocity ratio. This means that the efficiency increases and reaches its maximum value when the denominator approaches $M +$ velocity ratio. The maximum or limiting value of the efficiency is therefore

$$\text{limiting efficiency} = \frac{1}{M \times \text{velocity ratio}}$$

This variation is best illustrated by means of a graph plotting efficiency against load, as in figure 4.11.

Example 4.5

A screw jack with a single-start thread of 6 mm pitch is operated by a tommy bar whose effective radius is 57 mm; it can lift the following loads with the efforts shown.

Load (kg)	13.6	41	68	95	128	150
Effort (N)	9	22	33	49	62	76

Calculate the law of the machine and the limiting efficiency.

Solution First convert the loads into newtons; hence determine the mechanical advantage

Load (N)	133	402	667	990	1256	1540
Mechanical advantage	14.7	18.2	20.2	20.2	20.25	20.25

$$\text{Velocity ratio} = \frac{\text{distance moved by effort}}{\text{distance moved by load}}$$

$$= \frac{2\pi R}{\text{pitch}} = \frac{\pi D}{\text{pitch}}$$

$$= 59.6$$

Thus

mechanical efficiency	24.6	30.5	33.9	31.8	33.9	32.4

The law of the machine can be found by plotting the graph shown in figure 4.12. Limiting efficiency = $1/(M \times \text{velocity ratio}) = 1/(0.0474 \times 59.6) = 35.4\%$.

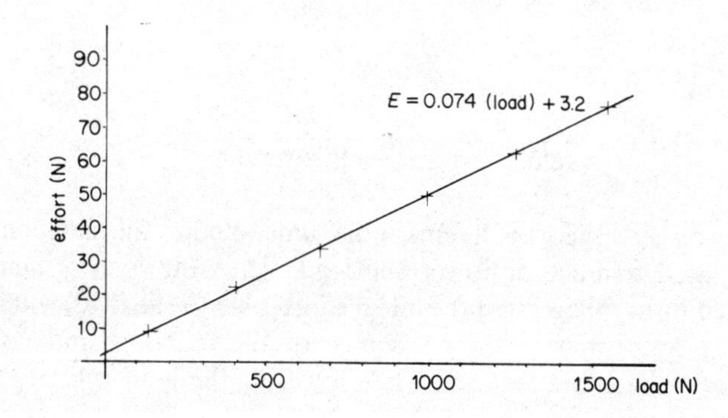

Figure 4.12

4.6 OVERHAULING IN A SIMPLE MACHINE—IDEAL MACHINES

A machine is said to 'overhaul' if, when the effort is removed, the friction and inertial resistances of the machine are insufficient to support the load, that is, the machine 'runs away' in the reverse direction to that of its normal operation. It is usual in lifting machines to design against this tendency by including ratchets and brakes which will prevent the run-back.

Now, let us consider the circumstances in which overhaul may occur.

Consider an ideal 'frictionless' machine, where the efficiency is 100 per cent. Since efficiency (%) = (mechanical advantage/velocity ratio) × 100

$$\text{ideal mechanical advantage} = \text{velocity ratio}$$

$$\text{ideal effort} = \frac{\text{load}}{\text{velocity ratio}}$$

All of this 'ideal' effort will be devoted to moving the load, since the ideal machine is frictionless and, without mass, it will have no resistance to motion.

In a real machine the actual effort will of course be greater than this, because it has to overcome the inertial and frictional resistances. We can regard this additional effort as the result of an additional load due to these resistances. Thus the actual effort equals ideal effort to overcome load plus effort to overcome resistances, that is

$$\text{effort, } E = \frac{\text{load}}{\text{velocity ratio}} + \frac{\text{resistance, } R}{\text{velocity ratio}}$$

But from the law of the machine (section 4.5) we know that

$E = M\ W + C$

thus

$$M\ W + C = \frac{W}{\text{velocity ratio}} + \frac{R}{\text{velocity ratio}}$$

The 'resistive' effort will remain the same whether the machine is being used to raise or lower the load; when the load is being lowered rather than lifted the 'ideal effort' (due to the load) will be tending to overcome the resistance of the machine and consequently the effort required when lowering the load will be

$$\text{friction effort} - \text{ideal effort} = M\ W + C - \frac{W}{\text{velocity ratio}} - \frac{W}{\text{velocity ratio}}$$

The machine will run away or overhaul when

$$M\ W + C > \frac{2\ W}{\text{velocity ratio}}$$

the critical case being when

$$M\ W + C = \frac{2\ W}{\text{velocity ratio}} \qquad (4.1)$$

Since efficiency = mechanical advantage/velocity ratio

$$\text{mechanical advantage} = \text{efficiency} \times \text{velocity ratio} = \frac{\text{load}}{\text{effort}}$$

$$\text{effort} = \frac{\text{load}}{\text{velocity ratio}} \times \frac{1}{\text{efficiency}} \qquad (4.2)$$

From equations 4.1 and 4.2

$$\frac{2\ W}{\text{velocity ratio}} = \frac{W}{\text{velocity ratio}} \times \frac{1}{\text{efficiency}}$$

therefore

$$\text{efficiency} = \tfrac{1}{2} = 0.5$$

Thus a machine whose efficiency is 50 per cent or greater will overhaul, that is, will run backwards when the effort is removed. This occurs because the frictional and inertial resistances of the machine are small enough to be overcome by the load.

TO THE STUDENT

At the end of this chapter you should be able to

(1) define the terms 'simple machine' (giving typical examples), 'mechanical advantage' and 'velocity (or distance or movement) ratio'
(2) appreciate that mechanical efficiency is the ratio of input energy to output energy and may be expressed as mechanical advantage/velocity ratio
(3) appreciate that mechanical advantage and hence mechanical efficiency of a simple machine will vary and know the factors which will produce this variation
(4) calculate the velocity or distance ratios for a variety of simple machines, given the design parameters
(5) solve simple problems involving load, effort, mechanical advantage, velocity ratio and efficiency
(6) obtain the necessary results from a test on a simple machine to enable the law of the machine to be determined in the form $E = M\ W + C$
(7) appreciate that the efficiency of a machine has a limiting value and be able to determine the value of this limiting efficiency
(8) know what is meant by 'overhauling' in a simple machine and appreciate the conditions under which overhauling will occur
(9) complete exercises 4.1 to 4.10.

EXERCISES

4.1 A mass of 120 kg is to be lifted using a pulley mechanism with two pulleys in the top block and one in the lower block. Determine the velocity ratio and mechanical advantage and hence the efficiency at this load if an effort of 500 N is required.

4.2 A load of 70 kg is lifted by means of a screw jack with a single-start square thread of 6 mm pitch, where the effort is applied at the end of a tommy bar giving an effective radius of 250 mm. If the efficiency of the screw jack at this load is 23 per cent, determine the effort required.

4.3 A crane is used to lift a load of 5×10^3 kg and at this load its efficiency is 75 per cent and its velocity ratio is 20. Determine the effort required.

4.4 A certain machine has a designed velocity ratio of 30:1 and it is found that it requires an effort of 2.5 kN to lift a mass of 5400 kg. What is the efficiency at this load?

4.5 A differential wheel and axle has drums 225, 300, 450 mm in diameter. If an effort of 200 N will lift a mass of 2500 kg, what will be the efficiency at this load?

4.6 A compound gear train, with a 25-tooth pinion engaging with a 100-tooth gearwheel, is mounted on the same axle as another pinion with 30 teeth meshing with a gear of 60 teeth on the output shaft. Determine the velocity ratio of this system and the mechanical advantage, if the efficiency is 80 per cent.

4.7 Tests on a block and tackle gave the following results.

Load (kg)	27	68	91	136	182
Effort (N)	31	57	75	101	128

By plotting an appropriate graph determine the law of the machine.

4.8 Tests on the lifting performance of a screw jack gave the following results.

Load (kg)	2.25	4.5	6.8	9.1	11.4	13.6
Effort (N)	2.2	3.0	4.2	5.28	6.4	7.5

The effort was applied at the end of an arm 100 mm long and the pitch of the single-start thread was 10 mm. By plotting an appropriate graph determine the law of the machine; hence determine the load that could be lifted by an effort of 10 N and calculate the efficiency at this load.

4.9 A simple lifting machine has a velocity ratio of 40 and it is found that to lift loads of 680 kg and 2000 kg it requires efforts of 350 N and 750 N respectively. Use these results to determine the law of the machine and hence predict the effort required to lift a load of 1500 kg.

4.10 A wheel and differential axle assembly has an effort wheel diameter of 250 mm and the larger load axle is 130 mm diameter. It is necessary that an effort of 88 N shall lift a mass of 80 kg. If the efficiency at this load is 80 per cent determine the diameter of the smaller load axle.

NUMERICAL SOLUTIONS

4.1 3; 2.354; 78.5%
4.2 11.4 N
4.3 330 N
4.4 70.6%
4.5 10.2%
4.6 8; 6.4
4.7 $E = 0.06\ W + 13$
4.8 $E = 0.0535\ W + 1.1$; 29.6%; 16.8 kg
4.9 $E = 0.33\ W + 144$; 599 N
4.10 85.2 mm

5 Velocity

The object of this chapter is to extend the student's knowledge of uniformly accelerated motion and to develop this knowledge to cover angular motion.

5.1 CONSTANT-VELOCITY AND UNIFORMLY ACCELERATED MOTION

You will already have considered the simple case of constant-velocity motion in a straight line, defining velocity as the rate of change of position, determined by the formula

$$\text{velocity} = \frac{\text{distance or displacement}}{\text{time taken}}$$

$$v = \frac{s}{t}$$

We must appreciate that in many cases, the velocity does not remain constant, but undergoes a change (in either magnitude or direction) as for instance, when starting from rest and accelerating to a given velocity. You will already have defined this acceleration as being the rate of change of velocity (with respect to time) given by the formula

$$\text{acceleration} = \frac{\text{change of velocity}}{\text{time taken}}$$

Note This change of velocity may be in either magnitude or direction, but at this stage in our studies we shall consider that this change of velocity involves a change in magnitude only.

$$\text{Acceleration} = \frac{\text{final velocity} - \text{initial velocity}}{\text{time taken}}$$

$$a = \frac{v - u}{t}$$

We shall assume that the rate of change of velocity, that is, the acceleration, remains uniform (constant) during the time taken for the change of velocity to occur. We shall see later that this means that any forces involved will also be constant.

The distance travelled during any period of uniform acceleration (that is, a period during which the velocity changes at a constant rate) may be calculated by considering the average velocity achieved during this same time interval.

$$\text{Distance} = \text{average velocity} \times \text{time}$$

$$s = \frac{(u + v)}{2}t$$

The final velocity v achieved after any given time t will depend on the initial velocity and on the uniform acceleration. This final velocity will vary as the time changes, since the initial velocity and the acceleration will both remain constant.

If this initial velocity is 5 m/s and the acceleration is 2 m/s^2, which means that the velocity will be increased by 2 m/s every second, then

after 1 s the velocity will be

$$v = 5 + 2 = 7 \text{ m/s}$$

after 2 s

$$v = 5 + (2 \times 2) = 9 \text{ m/s}$$

and after 3 s

$$v = 5 + (2 \times 3) = 11 \text{ m/s}$$

This can be expressed in general terms as

$$v = u + at$$

Example 5.1

A car starting from rest achieves a velocity of 45 km/h in 50 s. Find its acceleration. If this acceleration is maintained, find the time taken to reach a velocity of 55 km/h starting from rest. How far does the car travel in reaching this velocity?
Solution

$$45 \text{ km/h} = \frac{45 \times 10^3}{60 \times 60} = 12.5 \text{ m/s}$$

$$\text{Acceleration} = \frac{\text{change of velocity}}{\text{time}} = \frac{v-u}{t}$$

$$= \frac{12.5-0}{50} = 0.25 \text{ m/s}$$

$$55 \text{ km/h} = \frac{55 \times 10^3}{60 \times 60} = 15.2 \text{ m/s}$$

$$\text{Acceleration} = \frac{\text{change of velocity}}{\text{time}}$$

$$\text{Time} = \frac{\text{change of velocity}}{\text{acceleration}} = \frac{15.2-0}{0.25} = 60.8 \text{ s}$$

$$\text{Distance} = \frac{(u+v)}{2}t = \frac{(0+15.2)}{2}60.8 = 462.08 \text{ m}$$

From these important expressions involving the simple relationships between velocity, distance and acceleration for uniformly accelerated motion, that is

$$a = \frac{v-u}{t} \tag{5.1}$$

$$s = \frac{(u+v)}{2}t \tag{5.2}$$

$$v = u + at \tag{5.3}$$

we can build up other relationships and equations which will be useful in our studies. Thus substituting equation 5.3 into equation 5.2

$$s = \frac{u+(u+at)}{2}t$$

$$= \frac{(2u+at)}{2}t$$

thus

$$s = ut + \tfrac{1}{2}at^2 \tag{5.4}$$

This expression enables the distance travelled during a period of uniform acceleration to be more readily determined.

Also, from equation 5.2

$$u+v \text{ or } v+u = \frac{2s}{t} \tag{5.5}$$

and from equation 5.3

$$v - u = at \tag{5.6}$$

Multiplying equations 5.5 and 5.6 together

$$(v+u)(v-u) = \frac{2s}{t} \times at$$

or

$$v^2 - u^2 = 2as \tag{5.7}$$

Example 5.2

A body with an initial velocity of 3 m/s and unknown uniform acceleration travels 1800 m in 50 s. Determine the acceleration and maximum velocity.

Solution Given s, u, and t we require a and v. To find a from equation 5.4

$$a = \frac{2(s - ut)}{t^2}$$

$$= \frac{2(180 - 3 \times 50)}{50^2}$$

$$= \frac{3300}{2500} = 1.32 \text{ m/s}^2$$

To find v from equation 5.3

$$v = 3 + (1.32 \times 50)$$
$$= 69 \text{ m/s}$$

5.2 DISTANCE–TIME, VELOCITY–TIME AND ACCELERATION–TIME GRAPHS FOR UNIFORMLY ACCELERATED MOTION

The relationship between distance, velocity, acceleration and time for any type of motion can be represented graphically by means of distance–time, velocity–time and acceleration–time graphs, arranged as shown in figure 5.1. For the particular case under discussion—the case of uniformly accelerated motion—the shapes of the graphs will be as shown.

Since the acceleration will be constant throughout the time (by definition of the type of motion being considered) the acceleration–time graph will be a horizontal straight line.

Since the velocity will be increasing at a uniform rate (because of the constant acceleration) the velocity–time graph will be a linear one—a straight line with a constant slope.

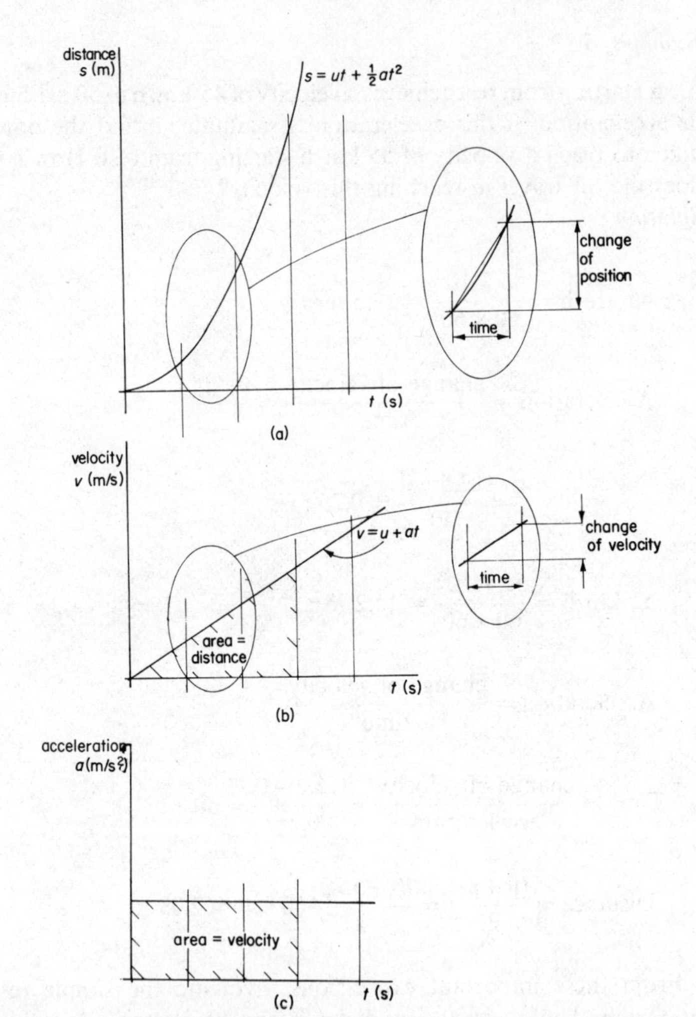

Figure 5.1 (a) Distance–time, (b) velocity–time, (c) acceleration–time graphs for uniformly accelerated motion, indicating the relationship between them

The average velocity during each second will increase, since the actual velocity is increasing, and consequently the distance travelled during each second will also increase at an increasing rate, making the distance–time graph a curve.

These deductions regarding the shapes of these graphs are confirmed by considering the equations determined in section 5.1. The relationship of distance to time was given by the equation

$$s = ut + \tfrac{1}{2}at^2$$

This is a quadratic expression (that is, the variable t appears raised to the second power) and the graph of any quadratic expression is a particular type of curve known as a parabola, which has the appearance of the curve shown.

The relationship of velocity to time was given by the equation

$$v = u + at$$

This is in the form of the standard expression for a straight line $y = mx + c$ (or $y = c + mx$), where c (or u) is the intercept on the vertical axis (the initial velocity) and m (or a) is the slope of the graph.

Furthermore the slopes and the areas of the graphs have a relationship to each other. The slope (gradient) of the velocity–time graph, which is constant throughout the time under consideration, may be calculated by constructing a right-angled triangle as shown and determining the tangent of the angle so formed by dividing the vertical height by the base (tangent = opposite/adjacent). The vertical height represents a change of velocity and the base represents a time interval t therefore the slope or gradient will have the units of

$$\frac{\text{change of velocity}}{\text{time}} = \text{acceleration}$$

Thus the slope of the velocity–time graph represents the acceleration.

Similarly if we consider the distance travelled during a particular time interval by constructing a right-angled triangle (figure 5.1a), considering the 'average slope' as before, the gradient is determined by dividing the vertical 'distance' ordinate by the horizontal 'time' ordinate giving the slope the units of

$$\frac{\text{distance}}{\text{time}} = \text{velocity}$$

Thus the slope of the distance–time graph represents the velocity.

We can see that the value of the gradient or slope of one graph will give the 'derived' quantity. In more advanced studies in mathematics this same relationship is used and applied when using the calculus technique of differentiation.

A further relationship can be observed when considering the areas beneath the graphs. The rectangular area underneath the acceleration–time graph is determined by multiplying the base (the time concerned) by the height (the acceleration). Considering the units involved we have

$$\text{time} \times \text{acceleration} = \text{s} \times \text{m/s}^2 = \text{m/s}$$

Thus the area under the acceleration–time graph represents the velocity.

This is confirmed if we consider the area under the acceleration–time graph (figure 5.1c) after 1 s and after 2 s. The area will obviously have been doubled and by considering the corresponding velocity–time graph for an object starting from rest, then the velocity after 2 s will be double its value after 1 s.

The triangular-shaped area under the velocity–time graph is determined by multiplying the base by the height (although this time a factor of $\tfrac{1}{2}$ is introduced since we are dealing with the area of triangle, which is half the area of the rectangle drawn on the same base to the same height). Thus

$$\text{area of triangle} = \tfrac{1}{2}(\text{base} \times \text{height})$$

which can be conveniently written

$$\text{area of triangle} = \text{base} \times \tfrac{1}{2}\text{height}$$

In this case the units of the base are time and the height measures velocity, while $\tfrac{1}{2}$height will give the average velocity during any particular time interval; thus area = time × average velocity

$$= s \times \frac{m}{s}$$

$$= m$$

Thus the area under the velocity–time graph represents the distance travelled.

Looking at the graphs we see that when the time interval is doubled the area underneath the curve is considerably more than doubled and this is reflected in the height (representing distance) under the distance–time graph at the corresponding times.

This relationship between curves and corresponding areas is used in more advanced studies of mathematics when dealing with the technique of integration.

These relationships between distance–time, velocity–time and acceleration – time graphs remain true for any type of motion. Practically, we can use these relationships when using Fletcher's trolley equipment in the laboratory to demonstrate Newton's law of motion.

5.3 UNIFORMLY ACCELERATED MOTION DUE TO GRAVITY

One of the most common applications of uniformly accelerated motion is that which takes place when an object is allowed to fall freely through space towards the Earth.

When two bodies are close to each other they experience a mutual attraction. The idea of this attractive force was first expounded by Sir Isaac Newton, an English scientist whose work laid the basis for much of today's scientific knowledge. He found that this attractive force was proportional to the product of the masses and inversely proportional to the square of the distance between them. This can be written as

$$\text{force} = \frac{(M_1 \times M_2)}{d^2} k$$

where M_1 and M_2 are the masses, distance d apart and k is a constant of proportionality.

All the objects that we use are close to a very large mass — the Earth — and consequently they experience an attractive force pulling them towards the Earth. This attractive force pulling objects towards the Earth is known as the gravitational force. This gravitational force acting on a body will cause the body to 'fall' towards the surface of the Earth and this movement will take place with uniform acceleration, since, following Newton's second law of motion (section 6.2), the application of a constant force will produce a constant acceleration. The force will remain constant although the object will move nearer to the centre of the Earth as it continues its fall; the variation in distance is minimal since the depth of any fall in our experience compared to the radius of the Earth is negligible. Because of the continuing effect of the gravitational force, and hence gravitational acceleration, the velocity (or speed) of the fall will tend to continue to increase until it reaches a limiting value. This is due to the resistance to motion it will experience by its movement through the air, which will increase as the velocity increases.

The acceleration due to gravity is the same for any object falling at the same place on the Earth.

The phenomenon of gravitational acceleration was first demonstrated by Galileo Galilei (1564–1642) an Italian physicist from Pisa.

The Earth is not a perfect sphere and consequently the distance from the surface of the Earth to its centre will vary depending on the geographical location. This variation in distance will cause a variation in the attractive force experienced by a body and, following Newton's laws of motion, (chapter 6), a variation in the gravitational acceleration will be produced. The diameter of the Earth is larger when measured across the Equator than when measured across the Poles. This variation due to locality is best illustrated by table 5.1.

Generally in the United Kingdom a figure of 9.81 m/s^2 is sufficiently accurate for most engineering purposes, denoted by the symbol g.

You may have noticed that the term 'gravitational attraction' or 'gravitational force' has been used to measure the force which pulls

Table 5.1

Locality	*Acceleration due to Gravity on Free-falling Body*
Equator	9.781
Pisa	9.806
London	9.812
Glasgow	9.816
North and South Poles	9.832

a body towards the Earth, whereas in everyday non-technical language the term 'weight' might have been used. This term 'weight' is very loosely used, sometimes implying a measure of the gravitational attraction acting on a body and sometimes implying a measure or comparison of the mass of a body, that is, the amount of substance contained in a body.

5.4 VELOCITY AS A VECTOR

You will know that velocity is the more technical term used by engineers (rather than the term 'speed') when they wish to describe how fast an object is moving. Engineers must be precise in their terminology; consequently when discussing motion they are not only concerned with how fast an object moves, they are also concerned with the direction in which it moves. They use the term 'displacement' rather than distance when discussing movement, since displacement will specify how far the object moves and will also specify the direction in which the movement takes place (for example, 300 mm at 36° W of N).

Velocity is the rate at which the displacement occurs, that is

$$\text{velocity} = \frac{\text{displacement}}{\text{time}}$$

Thus velocity will have magnitude and direction and both must be specified for its complete definition.

Any quantity which has both magnitude and direction is known as a *vector quantity* (section 2.1) since it can be represented in magnitude and direction by means of a straight line (a vector) whose length to some suitable and stated scale represents the magnitude of the quantity and whose direction indicates the direction in which the quantity is acting. Displacement and its derivative quantities velocity and acceleration are all vector quantities, as are force and momentum.

Representing velocities by means of vectors is a particular asset when we wish to add together two or more velocities. An object frequently has a total velocity which is the sum of two component velocities: for instance, an aircraft may be moving through the air with a certain airspeed in a certain direction but the air itself will be moving due to the wind, and consequently the actual movement of the aircraft relative to the Earth's surface will be the vector sum of these two velocities.

Suppose a light aircraft is travelling at 600 km/h on a course due north but at the same time the wind is blowing at 80 km/h in an easterly direction; assuming a scale of 1 mm representing 20 km/h, the vector representing the velocity of the aircraft (relative to the air) can be drawn in as in figure 5.2 (a line 30 mm long). On to this vector is added a vector (4 mm long) representing the wind velocity. The resultant or total velocity of the aircraft relative to the Earth can be determined by completing the triangle, as was done to find the resultant in the triangle of forces (section 2.2).

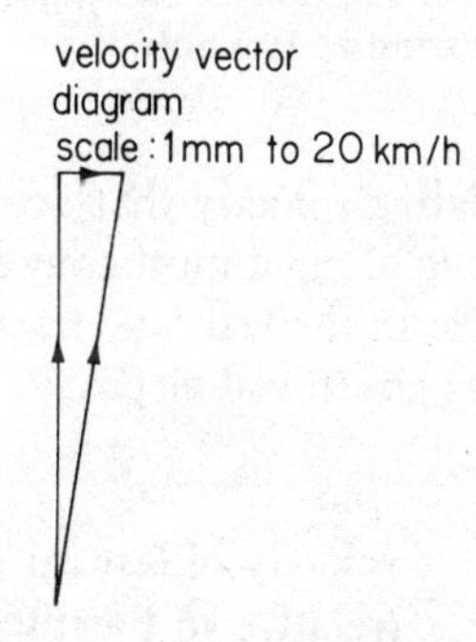

Figure 5.2 Velocity vector diagram for an aircraft travelling due north at 600 km/h with wind blowing due east at 80 km/h

A further example of this addition of velocities is the movement of a passenger on board a ship which is moving through the sea.

Example 5.3

A passenger strolls across the deck of a ship from the port side to the starboard side covering the distance of 50 m in 45 s. During this time the ship is moving through the sea at a velocity of 10 knots. Determine the resultant velocity of the passenger to the sea.

Solution This is solved by adding the vector representing the velocity of the passenger to the ship and the vector representing the velocity of the ship to the sea—see figure 5.3.

$$10 \text{ knots} = 18 \text{ km/h} = \frac{18 \times 10^3}{60 \times 60} = 5 \text{ m/s} = \text{velocity of ship to sea}$$

$$\frac{\text{dist}}{\text{time}} = \frac{50}{45} = 2 \text{ m/s} = \text{velocity of passenger to ship}$$

velocity of ship to sea

velocity of passenger to ship

resultant velocity of passenger to sea = 5.4 m/s

velocity vector diagram (scale 10 mm to 1m/s)

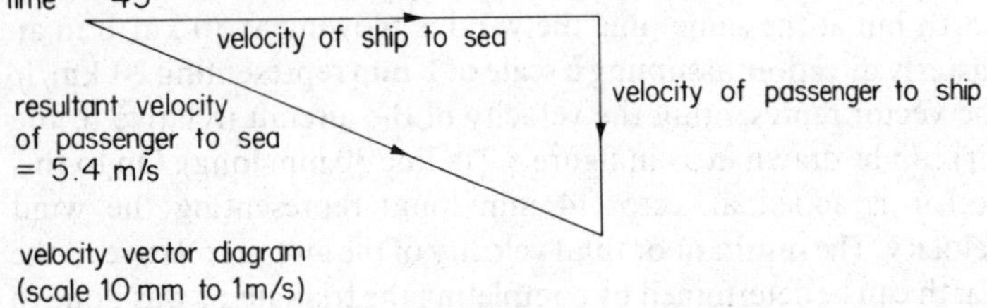

Figure 5.3 Velocity vector diagram for passenger moving 50 m across ship in 45 s when ship is moving at 10 knots

An example of relative velocity that occurs in everyday life is when a car is travelling along a motorway at, say, 50 mph, being overtaken by a vehicle in the fast lane travelling at 70 mph. The faster vehicle appears to be travelling past the slower vehicle at only 20 mph.

$$\text{Relative velocity of fast car to slow car} = \text{velocity of fast car (relative to Earth)} - \text{velocity of slow car (relative to Earth)}$$

This may be also written

$$\text{velocity of fast car to Earth} = \text{velocity of fast car to slow car} + \text{velocity of slow car to Earth}$$

$$_{E}V_{F} = {_{S}V_{F}} + {_{E}V_{S}}$$

Furthermore, this introduces the idea that the motion under consideration—linear motion, motion in a straight line—has, in fact, been said to be straight relative only to the surface of the Earth, which itself is part of a sphere. This means that motion which we have called linear is only a special case of motion in a circular path (angular motion) where the radius determining the motion is very large, being the radius of the earth. These two concepts are dealt with in more detail in the following sections.

Projectiles

An interesting example of the use of resolution of vectors is to consider the motion of a projectile.

Consider an object projected with a velocity v inclined in a direction θ to the horizontal surface (figure 5.4). This inclined velocity may be resolved into two components, one horizontal, $v\cos\theta$, and the other vertical, $v\sin\theta$. Immediately the object is released it will be subjected to two resistances.

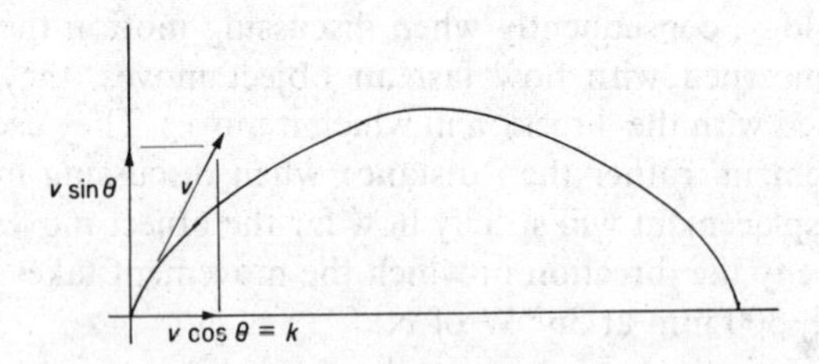

Figure 5.4

(1) The gravitational resistance which will oppose vertical motion and introduce a retardation of g (9.81 m/s^2).

(2) The resistance due to the air through which the object is

moving. However, when the object is moving comparatively slowly, this resistance may be neglected (wind resistance varies with the square of the velocity). This means that the horizontal component of the velocity remains constant.

As the object moves along it trajectory (path) it will tend to slow down due to the resistance in the vertical plane until it reaches its maximum height when the velocity in the vertical plane will be zero although in the horizontal plane it will still be moving with the same velocity.

Its vertical motion will have been

$v(\text{vert}) = v\sin\theta - gt$

and at its maximum height vertical velocity = 0

$$0 = v\sin\theta - gt$$

$$t = \frac{v\sin\theta}{g}$$

Applying the standard formula for uniformly accelerated motion

$$s = ut + \tfrac{1}{2}at^2$$

the maximum height may be found.

5.5 RELATIVE VELOCITY APPLIED TO MECHANISMS

We have already seen that all velocity is relative to some other velocity because ultimately all velocity on Earth is relative to the surface of the Earth, which has its own velocity.

This principle of relative velocity can be applied, on a much smaller scale, to mechanisms, to help determine the velocity and hence the acceleration of various points on the mechanisms. These are necessary steps in the process to determine forces set up in the parts of a mechanism.

The velocity of one moving point relative to a second moving point may be defined as the velocity of the first point when viewed by an observer imagined to be standing on and moving with the second point. This idea can be applied to any moving rigid link, which may be considered to make up any mechanism.

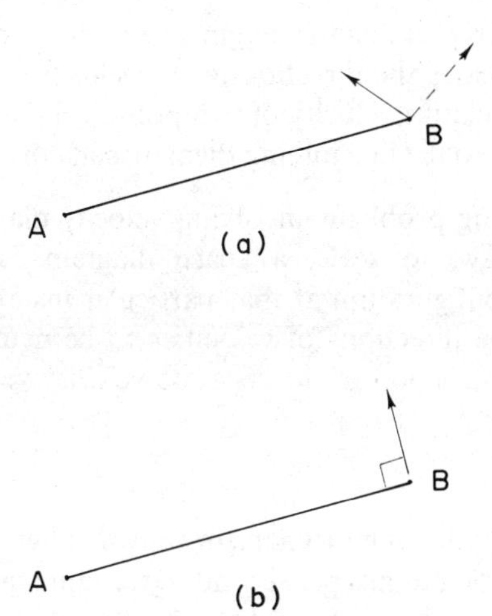

Figure 5.5 Movement of a rigid link. (a) B cannot move nearer to, or further away from, A without affecting the length of the link. (b) B maintains the same distance from A and can only rotate about A. Thus B's velocity relative to A must be tangential, that is, at 90° to the link. This is true whether A is moving or not

Consider the link AB in figure 5.5. Relative to the stationary framework, as represented by the paper, the ends of the link could have velocities differing widely in magnitude and direction. However, since the link must always retain its original dimensions, one end of the link cannot move nearer to, or further away from, the other end of the link, because this would cause an alteration of the length of the link. Thus the relative velocity of one end of the link to the other end must not have a component in the direction of the link, so the relative velocity can therefore only be at 90° to the axis of the link. Thus to an observer, imagined to be standing on one end of the link, the other end would appear to be rotating with an angular velocity (section 5.6) equal to the relative (linear) velocity divided by the length of the link.

In the construction of a velocity diagram we apply the following two principles.

(1) Velocity is a vector quantity and can be represented in magnitude and direction by a straight line whose length, to some

suitable scale, represents the magnitude of the velocity and whose direction indicates the direction of the velocity.

(2) The relative velocity of two points on the same rigid link must be at 90° to the line joining them (usually the axis of the link).

When solving problems involving velocity diagrams it is necessary to draw, to scale, a space diagram representing the mechanism configuration at that particular instant of time. This will enable the directions of velocities to be determined and so assist the construction of the separate velocity vector diagram.

Example 5.4

In the piston and crank mechanism shown in figure 5.6, the crank OA rotates at 50 radians per second. Determine the linear velocity of the piston for the configuration shown.

Solution Figure 5.6 shows the space diagram and velocity diagram, giving velocity of piston (B) as 254 m/s.

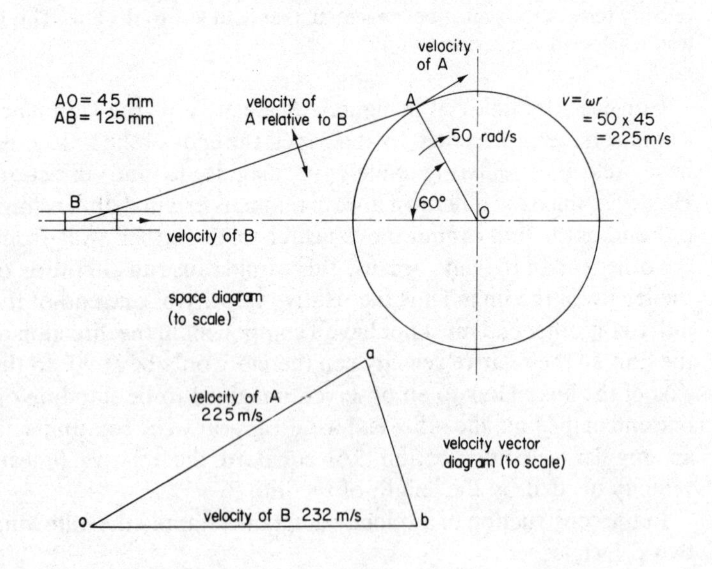

Figure 5.6 Piston and crank mechanism with corresponding velocity diagram

5.6 CIRCULAR MOTION AND ANGULAR VELOCITY

Many objects in everyday life and in various engineering fields move along a more complicated path (locus) than that of a straight line, but before dealing with such complex cases we should consider another type of motion of prime importance—angular motion, that is, motion in a circular path around a fixed centre. In engineering such a motion occurs in the case of gearwheels (non-epicyclic) pulley wheels and belt drives.

An essential difference between linear and angular motion is that in linear motion all particles of the body move with the same velocity, but in angular motion particles nearer to the centre of rotation move more slowly than those further away from the centre. (Consider the movement of different points along the length of the spoke of a bicycle wheel.)

When a body is rotating, the magnitude of the linear velocity of any point will depend on the radius of that point from the centre of rotation; the direction of the linear velocity will be changing all the time, but at any particular instant it can be said to be tangential (at 90°) to its radius.

In everyday terminology the angular speed of rotation is often referred to in terms of revolutions per minute (rpm), or more correctly in the SI system, in revolutions per second; the abbreviations for these quantities are rev/min and rev/s.

In more technical terminology we measure angular velocity using a system which enables a conversion between the speed of rotation and the corresponding linear velocity to be more readily carried out. This is particularly useful to a technician when discussing cutting speeds and belt speeds on pulleys. We divide each complete revolution, 360°, into 6.28 (2π) radians (a radian is defined as the angle subtended at the centre of a circle by an arc length equal to the radius of the circle). Thus

$$1 \text{ radian} = \frac{360°}{2\pi} = \frac{360}{6.28}$$

$$= 57.3°$$

This has the advantage that if we measure the angular distance in radians then the equivalent linear or arc distance will equal the number of radians multiplied by the radius. Thus

$$\text{arc distance} = \text{angular distance} \times \text{radius}$$

$$s = \theta \times r \tag{5.8}$$

(see figure 5.7).

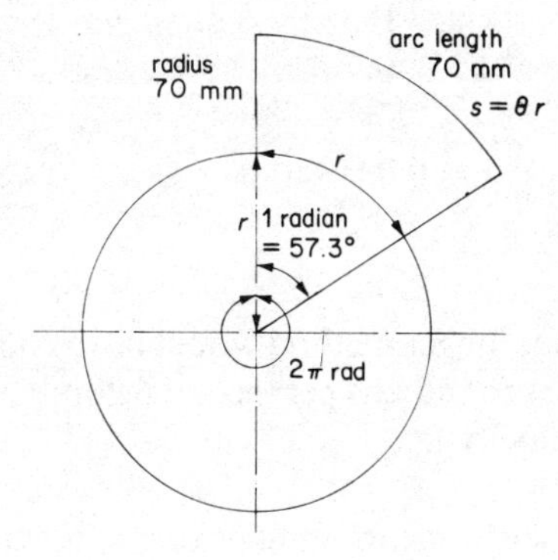

Figure 5.7 The relationship between arc length and radius is the radian measure

From our earlier studies we know that if we divide distance by time we obtain velocity and if we divide angular distance by time we obtain angular velocity.

$$\text{Angular velocity} = \frac{\text{angular distance (rad)}}{\text{time taken (s)}}$$

$$= \frac{\theta}{t} = \omega \text{ rad/s}$$

The symbol for angular velocity is omega, ω.

If we divide both sides of equation 5.8 by time, we obtain the relationship between linear and angular velocity, as follows. Dividing $s = \theta r$ by t

$$\frac{s}{t} = \frac{\theta r}{t}$$

but $\theta/t = \omega$, therefore

$$v = \omega r \tag{5.9}$$

thus

$$\text{linear velocity} = \text{angular velocity} \times \text{radius}$$

Example 5.5

A belt is running on a pulley 280 mm in diameter which is rotating at 1600 rev/min. Assuming no slip occurs between the belt and the pulley, calculate the belt speed in m/s.

Solution To convert rev/min to rad/s

$$1600 \text{ rev/min} = \frac{1600}{60} \times 2\pi \text{ rad/s}$$

$$\omega = 167.5 \text{ rad/s}$$

but

$$v = \omega r$$

$$= 167.5 \times \frac{0.280}{2} = 23.45 \text{ m/s}$$

Example 5.6

A racing-car moves around a circular track of 50 m radius at a speed of 120 km/h. Find the angular speed in radians per second.

Solution

$$120 \text{ km/h} = \frac{120 \times 10^3}{60 \times 60} = 33.33 \text{ m/s}$$

$v = \omega r$ therefore

$$\omega = \frac{v}{r} = \frac{33.33}{50} = 0.667 \text{ rad/s}$$

Let us now consider angular acceleration, which may be defined as the rate of change of angular velocity with time. The units involved may be found as follows.

$$\text{Angular acceleration, } \alpha = \frac{\text{change of angular velocity (rad/s)}}{\text{time (s)}}$$

$$= \text{rate of change of angular velocity}$$

$$= \frac{\omega}{\text{time}} \text{ rad/s}^2$$

We can establish the relationship between linear and angular acceleration by dividing both sides of equation 5.9 by time, that is, dividing $v = \omega r$ by t

$$\frac{v}{t} = \frac{\omega r}{t}$$

but $\alpha = \omega/t$, therefore

$$a = \alpha r \qquad (5.10)$$

linear acceleration = angular acceleration × radius

Example 5.7

A flywheel is accelerated from 50 rev/min to 90 rev/min in 45 s. Determine the angular acceleration in rad/s^2.
Solution

$$\text{Change in angular velocity} = 90 - 50$$
$$= 40 \text{ rev/min}$$

$$40 \text{ rev/min} = \frac{40}{60} \times 2\pi \text{ rad/s}$$

$$= 4.188 \text{ rad/s}$$

$$\text{angular acceleration} = \frac{\text{change of angular velocity}}{\text{time}}$$

$$= \frac{4.188}{45}$$

$$= 0.093 \text{ rad/s}^2$$

(An alternative but longer method would be to convert both initial and final velocities to radians per second before subtracting to find the change of velocity.)

It is perhaps worth mentioning at this point that even when a point is moving along a circular path with a constant angular velocity, the direction of its 'linear' velocity will be changing all the time, although the magnitude of its linear velocity may remain constant. Since velocity has both magnitude and direction and a change of velocity (an acceleration) can be brought about by a change in either direction or magnitude, the point moving along a circular path will be experiencing an acceleration at all times. This inward acting acceleration, known as *centripetal acceleration*, is the acceleration necessary to produce circular motion, as we shall see when considering Newton's first law of motion (chapter 6).

Example 5.8

A truck moving along a straight track accelerates from 0.1 m/s to 0.3 m/s in 15 s. Find (a) the linear acceleration, (b) the corresponding angular velocities of the wheels at the start and finish of acceleration if they are 70 mm in diameter, and (c) the angular acceleration of the wheels.

Solution (a)

$$\text{Linear acceleration} = \frac{\text{change of velocity}}{\text{time}}$$

$$= \frac{0.3 - 0.1}{15} = \frac{0.2}{15}$$

$$= 0.013 \text{ m/s}^2$$

(b) Angular velocity at 0.1 m/s

$$\frac{v}{r} = \frac{0.1}{0.035} = 2.857 \text{ rad/s}$$

Angular velocity at 0.3 m/s

$$\frac{v}{r} = \frac{0.3}{0.035} = 8.571 \text{ rad/s}$$

(c) $$\text{Angular acceleration} = \frac{\text{change of angular velocity}}{\text{time}}$$

$$= \frac{8.571 - 2.857}{15} = 0.381 \text{ rad/s}^2$$

TO THE STUDENT

At the end of this chapter you should be able to

(1) know the relationships between distance, velocity and acceleration for uniformly accelerated motion, both mathematically and graphically
(2) predict the behaviour of a body falling freely under the influence of the Earth's gravitational force
(3) appreciate that velocity is a vector quantity and use simple vector methods to solve problems involving relative velocities
(4) know the relationships between linear and angular distance, velocity and acceleration, and solve simple problems
(5) complete exercises 5.1 to 5.17.

EXERCISES

5.1 A motor cycle accelerates uniformly at 1.2 m/s^2. Calculate the time taken and the distance travelled before a speed of 80 km/h is reached from a standing start.

5.2 An electric train starting from rest accelerates uniformly to its running speed of 50 km/h in 0.25 min. Determine the acceleration and the distance travelled.

5.3 A vehicle travelling at 30 km/h is brought to rest within a distance of 7 m. Find the retardation and the time taken to bring the vehicle to rest.

5.4 Sketch the velocity–time graph to represent the journey of a car which starts from rest, accelerates uniformly for 5 s to reach a speed of 50 km/h, travels at this speed for 70 s and then is decelerated at 1 m/s^2 until coming to rest. Determine (a) the acceleration, (b) time taken for deceleration, (c) distance travelled during acceleration, (d) distance travelled at constant speed, (e) distance travelled during acceleration, and (f) total distance travelled.

5.5 An object passes a certain point with a speed of 6 m/s and 2 s later its velocity is found to be 18 m/s. Find the uniform acceleration, the distance travelled during these 2 s and the distance travelled during the next 5 s, assuming that it continues with the same uniform acceleration.

5.6 A cam follower has a lift of 45 mm and the profile of the cam is designed to give a uniform acceleration of 15 m/s^2 during the first 15 mm of lift, uniform velocity during the next 15 mm and uniform retardation during the remaining 15 mm. Draw the velocity–time graph for the motion and determine the uniform velocity and the total time taken.

5.7 A stone dropped down a vertical mine shaft hits the bottom after 4 s. Calculate the depth of the shaft and the final velocity.

5.8 A body is projected vertically upwards (against gravitational acceleration of 9.81 m/s^2) with an initial velocity of 30 m/s. Determine the maximum height reached and the time taken to hit the ground again.

5.9 An object is fired vertically upwards with a velocity of 40 m/s from the top of a tower 50 m high. Determine the maximum height reached and the time taken before the object reaches ground level.

5.10 An object is thrown vertically upwards and returns to the thrower after 3 s. Determine the maximum height attained and the initial velocity of throwing.

5.11 An oarsman sets out to row across a river 30 m wide flowing with velocity 0.25 m/s. He intends to row straight across the river with a velocity of 0.75 m/s. Determine the magnitude and direction of his resultant velocity.

5.12 The airspeed of an aircraft which is set to fly due south is 700 km/h but the wind is blowing from the north-west with a velocity of 50 km/h. Determine the magnitude and direction of the resultant velocity.

5.13 In the space diagrams shown in figure 5.8, determine the velocity of the pistons for the configurations shown.

5.14 In the linkage mechanisms shown in figure 5.9 determine the magnitude and direction of the velocity of the points B.

5.15 Determine the angular velocity in radians per second of the wheels of a car while the car is travelling with a linear velocity of 50 km/h if the effective diameter of the wheels is 0.6 m.

5.16 A pulley 350 mm diameter is revolving at 650 rev/min. Neglecting the effects of slip, determine the belt speed in m/s.

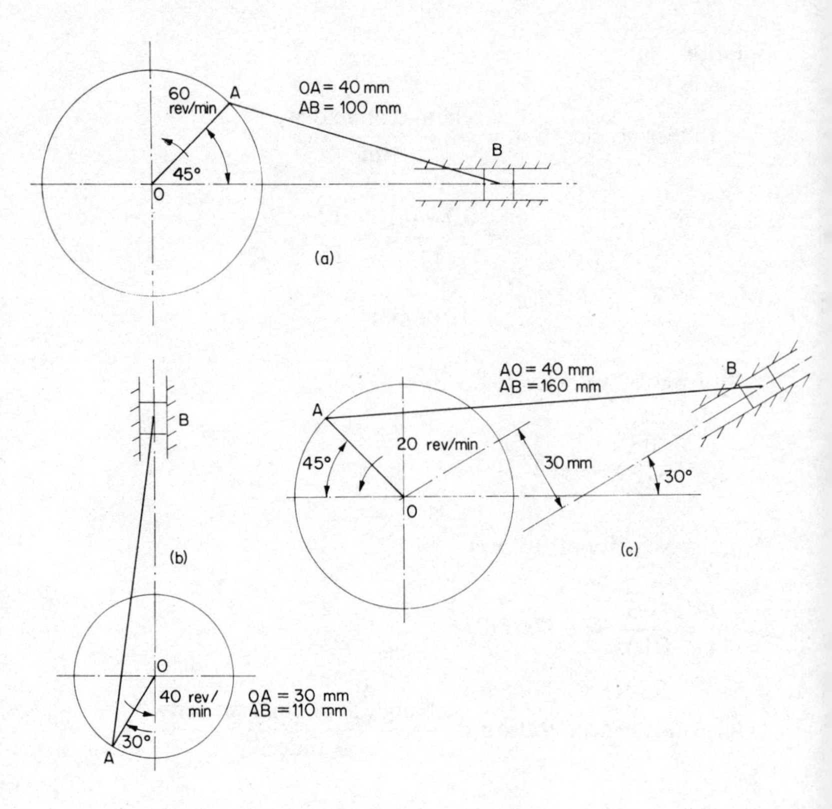

Figure 5.8

5.17 A truck whose road wheels are 450 mm in diameter is moving with a linear speed of 20 km/h when it is brought to rest within a period of 2.2 s. Determine the linear deceleration of the truck and the angular deceleration of the wheels.

NUMERICAL SOLUTIONS

5.1 18.3 s; 202 m
5.2 0.92 m/s^2; 103.5 m
5.3 4.96 m/s^2; 1.68 s

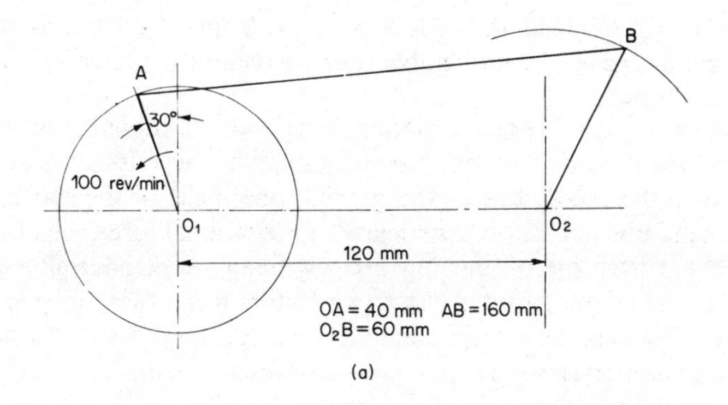

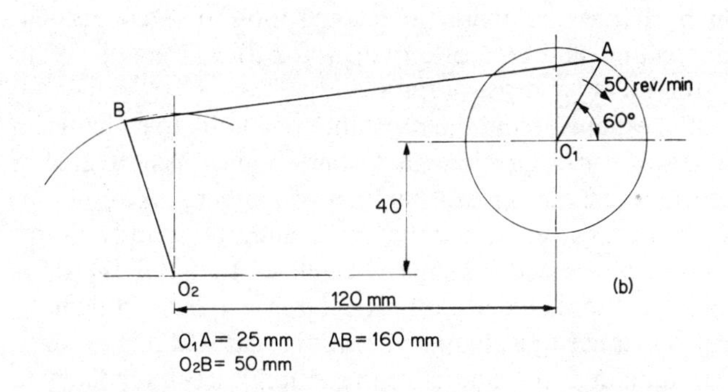

Figure 5.9

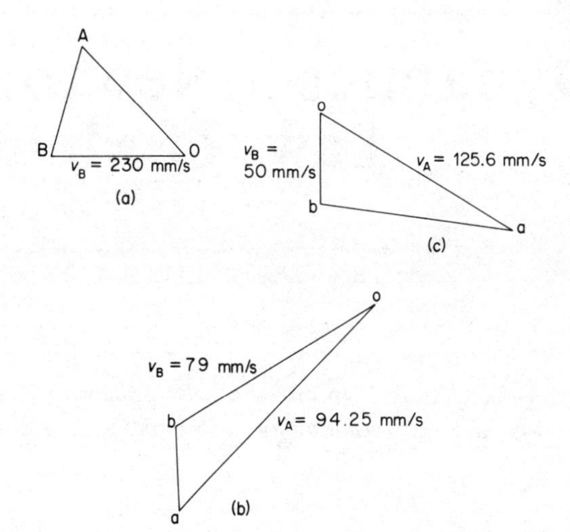

Figure 5.10

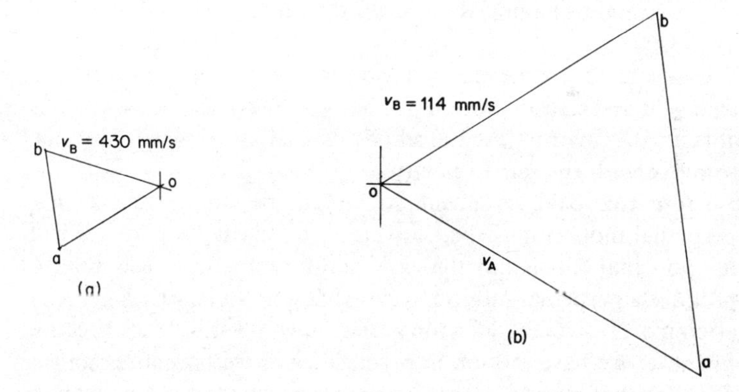

Figure 5.11

5.4 (a) 2.76 m/s^2, (b) 13.8 s, (c) 34.5 m, (d) 966 m, (e) 95.2 m, (f) 1095.7 m
5.5 6 m/s^2; 24 m; 540 m
5.6 21.2 m/s; 3.507 s
5.7 78.48 m; 39.24 m/s
5.8 45.9 m; 7.1 s
5.9 131.64 m; 9.25 s
5.10 11.02 m; 14.7 m/s
5.11 0.79 m/s at 71.5° to bank
5.12 111 km/h at 19° to S
5.13 (a) $v_B = 230$ mm/s, (b) $v_B = 79$ mm/s, (c) $v_B = 50$ mm/s—see figure 5.10
5.14 (a) $v_B = 430$ mm/s, (b) $v_B = 114$ mm/s—see figure 5.11
5.15 46 rad/s^2
5.16 11.9 m/s
5.17 2.52 m/s^2; 11.22 rad/s^2

6 Dynamics—Newton's Laws of Motion

The object of this chapter is to introduce the student to Newton's laws of motion and the terminology associated with a study of dynamics.

6.1 NEWTON'S FIRST LAW OF MOTION[1]

> A body will continue in its state of rest or of uniform motion in a straight line unless acted on by a force.

This means that some external force is necessary to bring about a change in the state of motion of a body. Otherwise a body which is at rest will remain at rest and a body which is moving will continue to move with the same velocity.

From this bald statement you might be led to believe that perpetual motion must be possible, but everyday experience will tell you that this is not the case—no inventor has been able to produce a perpetual-motion machine despite some very ingenious attempts. However, Newton's first law is still true because whenever we have motion in practical cases we have a resistance (force) to that motion (except in the case of certain motion through a vacuum). Reference has already been made to the frictional resistance to sliding motion, and frictional resistance also exists in rolling motion, although for a similar mass this latter resistance may be smaller than for the equivalent sliding mass. A moving body is also resisted by the material through which it moves. In the case of a body moving through air, the wind resistance will vary approximately as the square of the body's velocity: that is, if the velocity is doubled, the resistance is quadrupled; this means that streamlining is of considerable importance in the design of high-speed vehicles.

Not only does a body in motion experience a frictional resistance which will tend to modify the magnitude of the velocity, that is, reduce the speed, but in the gravitational field of the Earth, a moving body will also experience a gravitational force, which, in the case of a projectile moving through the air, may tend to alter the direction of the motion. These two factors mean that, in spite of everyday experiences apparently to the contrary, Newton's first law of motion is true and from this law we can obtain definitions of terms used in the study of dynamics.

Firstly, since a force is necessary to produce a change of motion, we can obtain our definition of a force—this being that 'a force is that which changes or tends to change motion'. Reference to Newton's second law of motion (section 6.2) will enable us to quantify the units to measure force.

Secondly, by considering the resistance of a body to a change in its motion, we can appreciate that some bodies have a greater resistance to a change in motion than do others. We call this resistance to a change in motion the *inertia* of the body and a simple consideration of everyday events will tell us that this inertia is proportional to the *mass* of the object. A large 'massive' object has a very great resistance to a change in motion, that is, if it is at rest it 'prefers' to remain at rest and once in motion it has a tendency to continue in that motion against any resistance.

We have already seen the importance of drawing a distinction between the mass and the weight (gravitational force) of an object and we are again grateful to Newton for his definition of mass as being 'the quantity of matter in a body'.

The distinction between mass (measured in kilograms) and weight (measured in newtons) is important and can be highlighted if we consider the methods by which we measure or compare masses—by using either a balancing device or a spring balance; these devices actually compare or measure the attractive (gravitational) force exerted on the object by the Earth.

Consider the use of the spring balance. We know that this attractive force will vary, following Newton's law of universal gravitation, according to the product of the masses and inversely with the square of the distance between them. We know that the Earth is not a perfect sphere and consequently if we take an object to different parts of the Earth—for example, if we take an object to the top of Mount Everest or down in a bathysphere to the bottom of the ocean—the distance to the centre of the Earth will vary and so the gravitational attractive force will vary (giving different readings on the spring balance) while the mass remains constant.

6.2 NEWTON'S SECOND LAW OF MOTION

> The applied force is proportional to the product of the mass and the acceleration produced.

This second law may be expressed in many different ways but the above version is the most useful format for engineers.

When engineers are concerned with the application of forces they are usually involved with the application of the force to a solid body. During the application of the force the mass of the body may be assumed to remain constant and thus Newton's second law may be expressed in such a form that the force is proportional to the acceleration produced. Thus

$$F \propto \text{mass} \times \text{acceleration}$$

or

$$F = k \times \text{mass} \times \text{acceleration} \qquad (6.1)$$

where k is a constant.

$$= K \times \text{acceleration}$$

where the constant $K = k \times \text{mass}$. We can use this proportionality relationship applied to the natural phenomenon of gravitational attraction to define a unit for the measurement of force.

We know that any mass will naturally have an acceleration, as a result of gravitational attraction, of 9.81 m/s^2. This is not a mathematically convenient value, so by taking equation 6.1 and choosing suitable units to ensure that the constant of proportionality, k, is unity we have: '1 unit of force acting on 1 unit of mass will produce 1 unit of acceleration'. Thus we have the definition of a newton, being that force which, when acting on a mass of 1 kg, will produce an acceleration of 1 m/s^2, that is

$$\text{force (N)} = \text{mass (kg)} \times \text{acceleration (m/s}^2\text{)}$$

This of course means that the gravitational force acting on a mass of 1 kg must be equal to 9.81 N (to produce the gravitational acceleration of 9.81 m/s^2). Similarly the gravitational force acting on any mass m is found by

$$\begin{aligned} \text{gravitational force} &= m \times 9.81 \\ &= mg \text{ N} \end{aligned}$$

where $g = 9.81\ \text{m/s}^2$. (Sometimes the term 'weight' is loosely used when referring to the gravitational force acting on a body.)

We can now apply our previous knowledge of acceleration to the engineer's version of Newton's second law. Thus

$$\text{force} = \text{mass} \times \text{acceleration}$$

but

$$\begin{aligned} \text{acceleration} &= \frac{\text{change of velocity}}{\text{time}} \\ &= \frac{v-u}{t} \end{aligned}$$

$$\text{force} = \frac{m(v-u)}{t}$$

Expanding this expression we have

$$\text{force} = \frac{mv - mu}{t} \qquad (6.2)$$

The numerator on the right-hand side of equation 6.2 consists of the difference of two terms, both of which are a product of mass and velocity. This product gives a quantity which is known as *momentum*, with units of kg m/s; this rather abstract quantity is sometimes referred to as the 'quantity of motion'. Momentum is a very useful concept when dealing with problems involving the transfer of motion between objects, particularly that which might occur in collisions. Then equation 6.2 becomes

$$\text{force} = \frac{\text{change of momentum}}{\text{time}}$$

or

$$\text{force} = \text{rate of change of momentum}$$

This is an alternative method of expressing Newton's second law which is useful to physicists.

Example 6.1

A car initially at rest accelerates uniformly reaching a speed of 50 km/h in 10 s. If the car and driver have a total mass of 750 kg, determine the accelerating force required.

Solution Convert to basic SI units

$$50 \text{ km/h} = \frac{50 \times 10^3}{3600} = 13.88 \text{ m/s}$$

$$\text{Acceleration} = \frac{\text{change of velocity}}{\text{time}}$$

$$= \frac{v - u}{t} = \frac{13.88}{10} = 1.388 \text{ m/s}^2$$

Newton's second law gives

$$\text{force} = \text{mass} \times \text{acceleration}$$

$$= 750 \times 1.388$$

$$= 1041.0 \text{ N}$$

Example 6.2

A vehicle initially travelling at 80 km/h has a shortest stopping distance of 60 m, conforming with the Highway Code. If the mass of the vehicle is 950 kg, determine the necessary braking force and the time taken to come to rest.

Solution Convert to basic SI units

$$80 \text{ km/h} = \frac{80 \times 10^3}{3600} = 22.2 \text{ m/s}$$

$$\text{Average velocity} = \frac{22.2 + 0}{2} = 11.1 \text{ m/s}$$

$$\text{Distance} = \text{average velocity} \times \text{time}$$

$$\text{Time} = \frac{60}{11.1} = 5.4 \text{ s}$$

$$\text{Acceleration} = \frac{\text{change of velocity}}{\text{time}}$$

$$= \frac{22.2}{5.4} = 4.11 \text{ m/s}^2$$

Alternative method of finding acceleration

$$v^2 = u^2 - 2as$$

$$a = \frac{u^2 - v^2}{2s} = \frac{(22.2)^2}{2 \times 60} = 4.107 \text{ m/s}^2$$

Newton's second law gives

$$\text{force} = \text{mass} \times \text{acceleration (or retardation)}$$

$$= 950 \times 4.1$$

$$= 3895 \text{ N}$$

Example 6.3

In an experiment it is found that a mass of 7 kg can just be pulled with a uniform velocity along a horizontal surface by an effort of

12 N acting horizontally. If an effort of 30 N replaces the 12 N, determine the acceleration and the coefficient of friction.

Solution Since a 12-N force is required to overcome friction

$$\text{friction force} = \mu mg$$

$$12 = \mu \times 7 \times 9.81$$

$$\mu = \frac{12}{7 \times 9.81} = 0.175$$

Force available to provide acceleration is

$$30 - 12 = 18 \text{ N}$$

$$F = m \times a$$

$$18 = 7 \times a$$

$$a = \frac{18}{7} = 2.57 \text{ m/s}^2$$

6.3 NEWTON'S THIRD LAW OF MOTION

For every force there is an equal and opposite reaction force.

This law means that if someone drives into the back of your car, you may take some satisfaction that your car will exert an equal and opposite force on his car.

A book resting on a table exerts a downward acting force equal to the gravitational attraction acting on it (its weight) on to the table. The table will exert an equal and opposite upward acting reaction on the book. Note that these action and reaction forces act on different bodies and consequently must not be confused with equal and opposite forces acting on a single body to produce equilibrium.

This effect can be readily seen if we consider what happens when we ride on a fairground machine and are whirled round in a circular path at great speed. Our bodies attempt to follow Newton's first law and move in a straight line, that is, attempt to move tangentially off the machine. The machine compels us to move in a circular path, thus changing our direction of motion and so subjecting us to an acceleration. This acceleration, which is always directed towards the centre of rotation, is known as the *centripetal acceleration* and in order to produce this acceleration there must (according to Newton's second law) be a force, acting in the direction of the acceleration. Therefore the carriage of the machine exerts an inward acting force on our bodies known as the *centripetal force* to ensure that we continue to move in a circular path. On the other hand, in attempting to continue motion in a straight line, our bodies exert an outward acting 'reaction' force on the carriage of the machine, known as the *centrifugal force*.

The centripetal and centrifugal forces are equal and opposite and are an example of the action and reaction forces referred to in Newton's third law.

6.4 FORCES REQUIRED TO PRODUCE AND MAINTAIN MOTION

We can now begin to apply Newton's laws of motion to problems associated with objects in motion.

First let us consider the case of uniform-velocity (constant-speed) motion. To maintain this motion it is usually necessary to overcome the frictional resistance and maintain the motion. This frictional resistance will often depend on the mass of the object (and may be expressed as so many newtons of resistance per kg mass). In the case of horizontal dry sliding friction, the resistance may be calculated from the formula $F = \mu mg$ (section 3.1). In the case of wind resistance the resisting force may be taken as varying with the square of the velocity (section 6.1).

If the applied force is less than the resistance, then the force will be insufficient to maintain the motion and the object will slow down (decelerate) at a rate which will depend on the difference between the applied force and the resistive force and on Newton's second law. Thus

$$\text{Resistance} - \text{applied force} = \text{mass} \times \text{deceleration}$$

For constant-velocity motion (no acceleration or deceleration) then

resistance = applied force

If we wish to produce an increase in the velocity (an acceleration) then an additional force is necessary ($F = ma$); then

total force = resistive force + accelerating force

So far we have considered motion along a horizontal plane, but if we consider motion in a plane other than horizontal we have to consider an additional force to deal with the effects of gravity, as follows.

Consider an object on an inclined plane. The gravitational force mg acting vertically downwards may be resolved into two components, one acting at 90° to the inclined plane, $mg \cos \theta$, and one acting parallel to the surface, $mg \sin \theta$ (figure 6.1a). This latter force ($mg \sin \theta$) will tend to produce motion down the plane against the frictional resistance. In the case of the dry sliding motion the frictional resistance on the inclined plane will be slightly less than that on a horizontal surface, since now only $mg \cos \theta$ will be acting normal to the surface, rather than mg as it would be on a horizontal surface.

To maintain the position on the slope a force equal to the difference between the frictional resistance and $mg \sin \theta$ will be necessary.

To produce motion up the slope it will be necessary to overcome the frictional resistance ($\mu mg \cos \theta$ in the case of dry sliding motion) plus the gravitational effects ($mg \sin \theta$). Should an acceleration up the slope be required, then a total force consisting of the sum of the frictional force, the accelerating force and the force necessary to overcome the gravitational effect is necessary.

In certain cases the line of action of the applied force is not parallel to the inclined plane. The most common example of this is when a horizontal force is used to move an object up an inclined

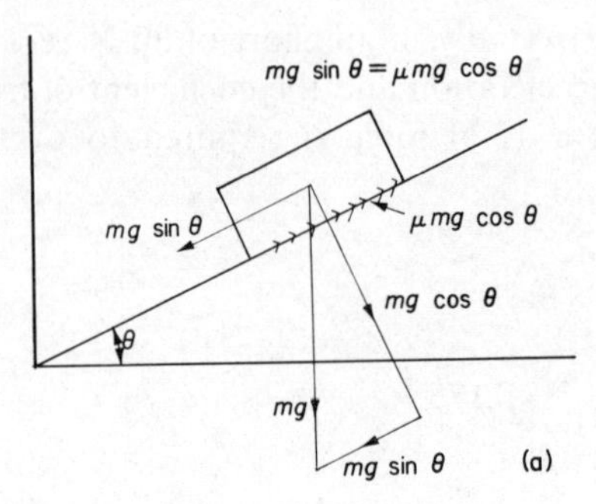

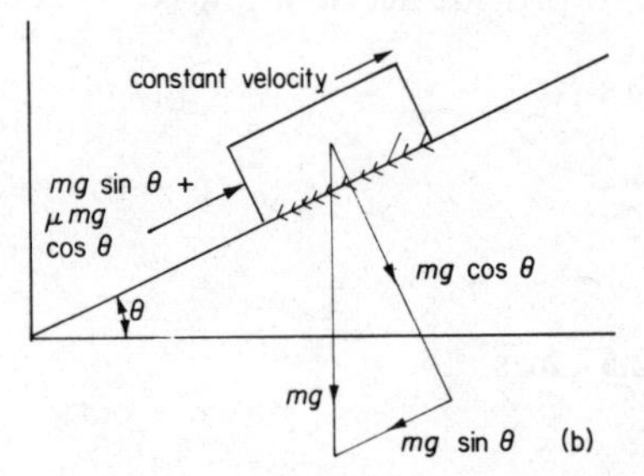

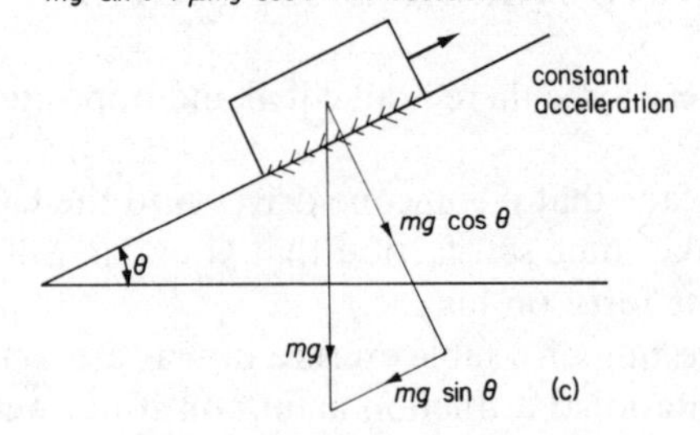

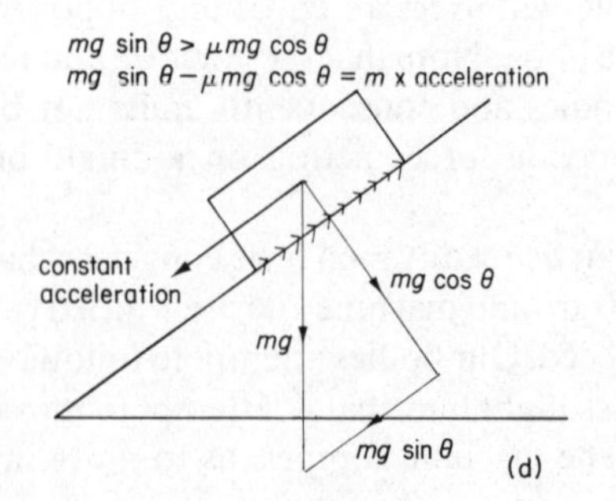

Figure 6.1 (a) Forces and reactions on a stationary mass on an inclined plane; (b) forces necessary to produce uniform-velocity motion up the incline; (c) forces necessary to produce accelerating motion up the incline; (d) forces acting when the mass slides down the incline

plane. In such a case it is necessary to apply a larger force than would be necessary if the force were applied parallel to the plane. The increased horizontal force is represented by the hypotenuse of a right-angled triangle whose adjacent side is the force acting parallel to the plane (see example 6.8).

This sort of problem is encountered when dealing with the turning moments applied to screw threads.

Example 6.4

Find the total force necessary to move twelve components, each of mass 3.25 kg, up a conveyor inclined at 25° to the horizontal. Assume that the force is to be applied parallel to the surface, that the components move with constant velocity and that the frictional resistance is 1.2 N per kg.

Solution

$$\text{Total mass} = 12 \times 3.25 = 39 \text{ kg}$$

$$\text{Total frictional resistance} = \text{mass (kg)} \times 1.2 \text{ (N/kg)} = 39 \times 1.2 = 46.8 \text{ N}$$

$$\begin{array}{l}\text{Force required to overcome} \\ \text{gravitational effects}\end{array} = mg \sin \theta$$

$$= 39 \times 9.81 \times \sin 25$$
$$= 39 \times 9.81 \times 0.4226$$
$$= 161.7 \text{ N}$$

No acceleration is required therefore

$$\text{total force} = \text{friction force} + \text{gravitational component} = 46.8 + 161.7 = 208.5 \text{ N}$$

Example 6.5

A cage of mass 4000 kg is being pulled up an incline of 1 in 35 against a frictional resistance of 0.3 N per kg. Determine (a) the force in the rope (pulling parallel to the incline) when the cage is accelerating at 0.6 m/s^2 and (b) the force when the cage is travelling at a constant velocity.

Solution

$$\text{Total frictional resistance} = 0.3 \times 4000 = 1200 \text{ N}$$

$$\begin{array}{l}\text{Force required to overcome} \\ \text{gravitational effects}\end{array} = mg \sin \theta$$

(35 m are travelled up the slope for a vertical rise of 1 m, thus hypotenuse = 35, opposite = 1 and $\sin \theta = 1/35$)

$$= 4000 \times 9.81 \times \frac{1}{35}$$
$$= 1121 \text{ N}$$

$$\begin{array}{l}\text{Force required to} \\ \text{produce acceleration}\end{array} = m \times \text{acceleration}$$

$$= 4000 \times 0.6$$
$$= 2400 \text{ N}$$

When cage is accelerating

$$\begin{array}{l}\text{total force} \\ \text{required}\end{array} = \begin{array}{l}\text{friction} \\ \text{force}\end{array} + \begin{array}{l}\text{gravitational} \\ \text{effort}\end{array} + \begin{array}{l}\text{accelerating} \\ \text{force}\end{array}$$

$$= 1200 + 1121 + 2400$$
$$= 4721 \text{ N}$$

When travelling at constant velocity

$$\text{total force} = 1200 + 1121 = 2321 \text{ N}$$

Example 6.6

If an object of mass 8 kg is placed on an incline of 50° and the coefficient of friction is 0.1, determine whether or not the object will move down the slope; if it will, specify the type of motion produced.

Solution

Gravitational component tending to produce motion down the slope $= mg \sin \theta$

$$= 8 \times 9.81 \times 0.7660$$
$$= 60.1 \text{ N}$$

Resistance to motion due to friction $= \mu mg \cos \theta$

$$= 0.1 \times 8 \times 9.81 \times 0.6428$$
$$= 5.045 \text{ N}$$

Force tending to produce motion is greater than frictional resistance therefore motion will occur.

Force available to produce acceleration $= 60.1 - 5.045$
$= 55.055$ N

Force = mass × acceleration, therefore

$$55.055 = 8 \times a$$
$$a = \frac{55.055}{8} = 6.88 \text{ m/s}^2$$

The object will move down the slope with an acceleration of 6.88 m/s^2.

Example 6.7

A load of 4 tonnes (t) is being hauled up an incline of 1 in 100 against frictional resistances of 350 N/t. If the load has an acceleration of 0.6 m/s^2, find the force necessary, assuming that the force is applied parallel to the incline.
Solution

Total frictional resistance $= 4 \times 350$
$= 1400$ N

Accelerating force = mass × acceleration
$= 4000 \times 0.6 = 2400$ N

Force to overcome gravitational effects $= mg \sin \theta$

$$= 4000 \times 9.81 \times \frac{1}{100}$$
$$= 392.4 \text{ N}$$

Total force required $= 1400 + 2400 + 392.4$
$= 4192.4$ N

Example 6.8

A loaded trolley of mass 150 kg is hauled up an incline of 1 in 8 by a wire rope which is horizontal when the pulling force is applied. The truck moves with a constant velocity against frictional resistances of 70 N. Find the force in the horizontal rope.
Solution Consider forces parallel to the inclined plane

force to overcome gravitational effects $= mg \sin \theta$

$$= 150 \times 9.81 \times \frac{1}{8}$$
$$= 184 \text{ N}$$

Total force parallel to plane $= 184 + 70 = 254$ N
($\sin \theta = 0.125$, $\theta = 7°$)

$$\text{Force acting horizontally} = \frac{254}{\cos 7} = 256 \text{ N}$$

1 We have already referred to the work of Sir Isaac Newton (1642–1727) in mentioning his law of universal attraction. This remarkable English scientist and mathematician was born on Christmas Day 1642, son of the Lord of the Manor at Woolsthorpe near Grantham. After domestic circumstances had forced him unsuccessfully to attempt farming he was sent to study at Trinity College Cambridge, where he took an interest in geometry and astrology.

In 1664–5 he, like all students, was sent down from Cambridge during the great plague and found time to contemplate and formed the basis for much of his later work in mathematics and mechanics.

He became the Lucas Professor of Mathematics and during the

next twenty years he worked on his *Principia* (published in 1687) in which he laid down the laws which are now known as Newton's laws of motion. In this he developed ideas which had been put forward by earlier scientists (such as Aristotle and Galileo) but because of his superior mathematical ability he was able to quantify their work. These laws formed the basis for 'Newtonian mechanics' which held true for over two hundred years before they were modified by the work of Einstein and others.

Newton's mathematical ability enabled him to develop the binomial theorem and differential and integral calculus; he also formulated several economic theories.

In 1689 he became M.P. for Cambridge University and in 1696 he became Warden and later Master of the Royal Mint. From 1703 until his death he was President of the Royal Society. In the SI system of units his name is used as the unit of force, where 1 newton is defined as force which, when applied to a mass of 1 kilogram, will produce an acceleration of 1 m/s^2.

TO THE STUDENT

At the end of this chapter you should be able to

(1) state Newton's three laws of motion
(2) define the terms 'force', 'inertia', 'momentum' and 'newton'
(3) solve problems involving force, mass and acceleration
(4) calculate the force necessary to produce constant velocity or uniformly accelerating or decelerating motion on a horizontal or inclined surface
(5) complete exercises 6.1 to 6.7.

EXERCISES

6.1 A vehicle of mass 950 kg travelling at 80 km/h is brought to rest in a distance of 87 m. Determine the time taken and the braking force required.

6.2 A machine tool has a mass of 275 kg. Find the total force necessary to accelerate it from rest to its cutting speed of 35 m/min in 0.8 s, if the cutting force plus frictional resistance are equal to 900 N.

6.3 A body of mass 2500 kg must accelerate from rest to a velocity of 60 km/h within (a) 30 s, (b) 250 m. Neglecting frictional resistances and assuming a horizontal surface, determine the necessary forces.

6.4 A vehicle of mass 1500 kg is moving up a slope of 1 in 15 against frictional resistances of 200 N per 1000 kg. Determine the driving force necessary if the vehicle is to move with an acceleration of 0.25 m/s^2.

6.5 A container of weight 2500 N is being pulled up a slope of 20° by a rope whose maximum tensile load is 3.5 kN. If μ between the slope and the container is 0.2, determine the maximum acceleration with which the container can be pulled up the slope.

6.6 A truck of mass 4500 kg is at rest on a slope of 1 in 12 when its brakes are released. If the frictional resistances are 350 N per 1000 kg, specify the motion, if any, of the truck.

6.7 The coefficient of friction between a body of mass 1500 kg and a surface is 0.35. If the body is at rest on a slope of 1 in 30, determine the force necessary to move the body down the slope with constant velocity.

NUMERICAL SOLUTIONS

6.1 7.83 s; 2.7 kN
6.2 1100 N
6.3 (a) 1389 N, (b) 1377 N
6.4 1656 N
6.5 5.179 m/s^2
6.6 Truck will move down slope with acceleration 0.467 m/s^2
6.7 34.7 N

7 Dynamics—Work and Energy

The aim of this chapter is to extend the student's knowledge of dynamics and to enable him to solve problems involving work energy and power.

7.1 WORK AND ENERGY

We have seen that the most common effect of a force is to produce motion and so, when we apply a force to an object, the object is likely to change its position, and to maintain the application of the force it will be necessary to move the force through a distance. This introduces us to a particular branch of study—that of dynamics, the study of forces in motion.

When we move the point of application of a force through a distance we say that we have done *work*. In everyday language the term 'work' is applied to many activities, including mental activity like studying. In technical terms we are more precise and work is always taken to refer to a force being applied through a distance, for example, in lifting an object against gravitational attraction, in pushing a body against frictional resistance, in applying a cutting force through a distance on a shaping machine. It should also be noted that work is also done when turning on a lathe; although the tool remains stationary the workpiece is revolving, thus creating relative motion between work and tool; consequently the force between work and tool can be said to move through a distance.

The unit used to measure work done is called the joule,[1] symbol J. One joule is equal to the work done when a force of 1 newton is moved through a distance of 1 metre (1 J = 1 N m).

Anything which has the ability to do work is said to possess energy. In everyday terminology we say that at the beginning of the day we are energetic: we have energy after a good night's sleep and are capable of doing work; at the end of the day we have less energy and are capable of doing less work. Obviously the units of work and of energy will be the same.

The human body is a machine which obtains its supply of energy from the chemical energy contained in the food it consumes. The considerations used to determine the efficiency of the body are complicated, but it seems that an efficiency of 20 to 25 per cent is achieved. The food consumed to provide energy—mainly carbohydrates and fats—originates from plant life which itself obtains energy from the Sun (as did the compressed and decomposed plants which form fossil fuels). Protein, used for body-building and repair, is obtained inefficiently and expensively from animals; alternative methods of protein supply are being devised.

We have already seen that energy may exist in many different forms, that machines are devices to convert energy and work done from one form to another (section 4.1) and that this conversion process must be carried out as efficiently as possible.

Efficiency is of particular importance when the cost of energy is so high and we are now realising that the world's energy resources are not unlimited (section 4.1). The use of the SI system of units is of considerable advantage in considering the conversion of energy from one form to another since, with this system, the same units are used for work and energy in different forms (such as mechanical and electrical) thus avoiding the use of conversion factors. (Thus when converting between heat and mechanical energy it is no longer necessary to refer to Joule's mechanical equivalent of heat.)

Example 7.1

Find the work done per stroke on a shaping machine when the length of the stroke is 40 mm and the cutting force is 90 N.

Solution

$$\text{Work done} = \text{force} \times \text{distance}$$
$$= 90 \times 40$$
$$= 3600 \text{ N mm} = 3.6 \text{ N m or J}$$

Example 7.2

Find the work done by pushing a mass of 700 kg through a distance of 4.5 m along a horizontal surface if μ is 0.2. Assume the force is applied horizontally.
Solution

$$\text{Friction force} = \mu mg = 0.2 \times 700 \times 9.81 = 1373.4 \text{ N}$$

$$\text{Work done} = \text{force} \times \text{distance} = 1373.4 \times 4.5 = 6180.3 \text{ N m or J}$$

Example 7.3

Find the work done in lifting a mass of 100 kg through a distance of 15 m.
Solution

$$\begin{aligned}\text{Work done} &= \text{force (against gravity)} \times \text{distance}\\ &= mgh\\ &= 100 \times 9.81 \times 15\\ &= 14715 \text{ N m} = 14.7 \text{ kJ}\end{aligned}$$

Example 7.4

A bar 70 mm in diameter is being turned on a lathe while the cutting speed is 120 rev/min. If the force on the cutting tool is 800 N find the work done per revolution and the work done per minute.
Solution

$$\begin{aligned}\text{Work done per rev} &= \text{force} \times \text{distance}\\ &= 800(\pi \times 70)\\ &= 176000 \text{ N mm}\\ &= 176 \text{ N m or J}\end{aligned}$$

$$\begin{aligned}\text{Work done per min} &= 176 \times 120 \text{ J}\\ &= 21.12 \text{ kJ}\end{aligned}$$

7.2 WORK DONE BY A CONSTANT FORCE

The mechanical work done when a constant force is moved through a distance can be shown graphically by plotting a graph of force vertically against distance horizontally (figure 7.1). The area underneath the graph is determined by multiplying the force (vertical height) by the distance (horizontal distance) and represents the work done.

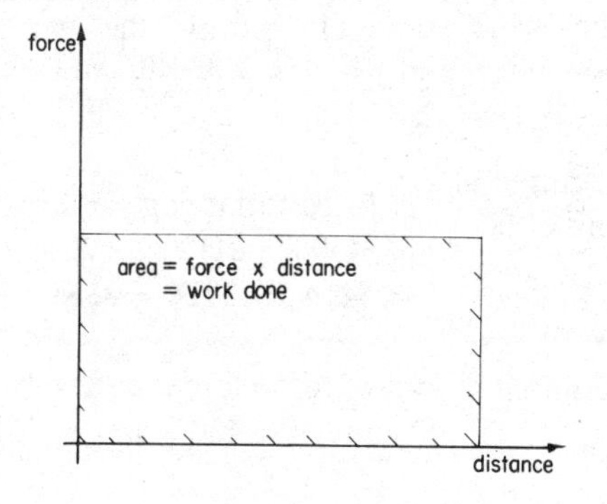

Figure 7.1 Work-done diagram representing the work done by a constant force

7.3 WORK DONE BY A VARYING FORCE

In many practical cases where a force is applied through a distance, the force does not remain constant during its application. For instance, when stretching or compressing a spring the force required will increase linearly as the compression or extension away from the free length increases.

Another example of a linearly increasing force that we have already met is that which occurs during the tensile or compression test. In other cases the force may vary cyclically or irregularly during the movement, depending on the resistance to motion. In such cases it is necessary to consider the average force applied and to multiply this by the distance travelled.

The graphical technique mentioned previously can be used to great advantage since the pattern of the variation of the force may readily be seen and possibly the average value more readily determined. The area under the graph will represent the work done as before. In the case of a linearly varying force the area will be a trapezium.

Example 7.5

A lift cage of mass 450 kg has to be hauled through a vertical height of 200 m. The supporting rope has a mass of 2 kg per metre. Calculate the work done in lifting the cage. Draw the force – distance graph and use this to verify your answer.

Solution (See figure 7.2.)

Force required at beginning of lift = force to lift cage and rope

$= (450 \times 9.81) + (2 \times 200)\ 9.81$

$= 4414.5 + 3924 = 8338.5$ N

Force required at end of lift = force to lift cage

$= 450 \times 9.81 = 4414.5$ N

$$\text{Average force} = \frac{8338.5 + 4414.5}{2} = 6376.5 \text{ N}$$

Work done = force × distance

$= 6376.5 \times 200$

$= 1\,275\,300$ N m $= 1.275$ MJ

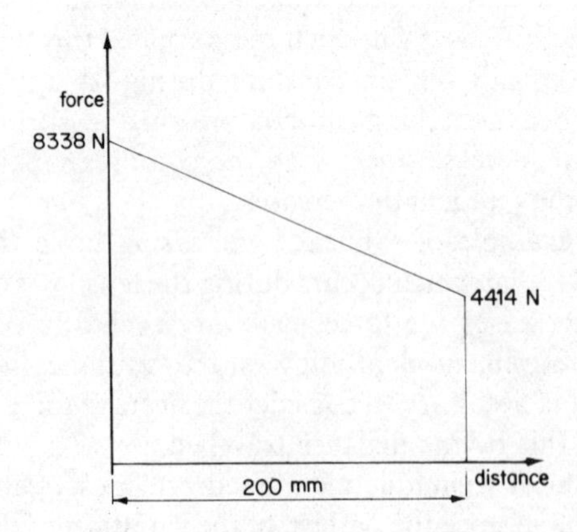

Figure 7.2

7.4 WORK DONE BY AN INCLINED FORCE

Again in many practical situations it is not possible to apply a force in the direction in which the motion is to occur. An example of this is when a wheelbarrow is pushed or pulled along the ground. The applied force is directed downwards towards the surface of the ground, while the motion is along the surface (figure 7.3). Here it is necessary to resolve the force into two components, one acting along the direction of the motion and the other in a direction perpendicular to the motion. This latter component may affect the resistance to motion (section 3.1).

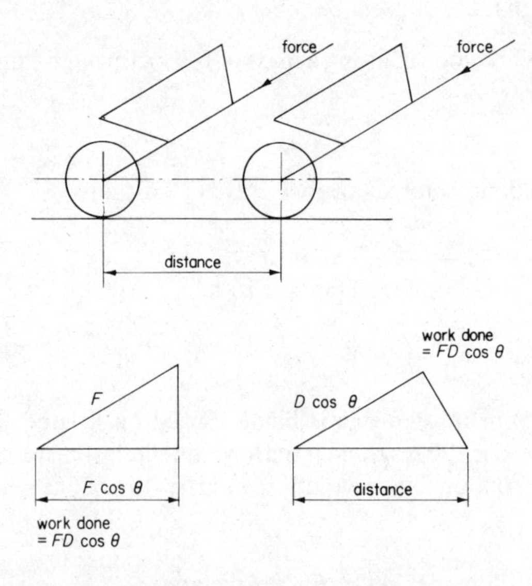

Figure 7.3 The force exerted in pushing a wheelbarrow acts at an angle to the direction in which the motion is produced

Example 7.6

A loaded truck is moved by a force of 600 N applied at an angle of 25° to the horizontal surface over which the truck is moving. Determine the work done in moving the truck a distance of 750 m.

Solution

$$\text{Horizontal component of force} = 600 \cos 25^\circ = 543.8 \text{ N}$$

$$\begin{aligned}\text{Work done} &= \text{force} \times \text{distance} \\ &= 543.8 \times 750 \\ &= 407\,850 \text{ N m} = 407.9 \text{ kJ}\end{aligned}$$

7.5 MECHANICAL ENERGY

We have seen that the application of a force usually produces motion but in those cases where the resistance to motion is greater than the applied force, a deformation is produced (section 1.2).

When movement is produced which is against the gravitational resistance, then the work done in lifting an object to its new height is stored within the object as energy while it remains in this position (since energy cannot be created or destroyed). The energy possessed by the body by virtue of its position is known as *potential energy*.

When a force is applied to an object at rest and motion is produced, then the work done in producing the motion is stored within that body while it is moving at that velocity. This energy due to motion, possessed by all moving objects, is known as *kinetic energy*.

In these two examples of mechanical energy the distance through which the force is moved is usually considerable, but in cases where a deformation is produced the force only moves through a distance equal to the deformation. Except in cases involving springs this distance will be quite small. Furthermore the force producing the deformation usually varies as the deformation increases (as in the case of the tensile test, section 1.5). The work done by the force (generally a variable force) in producing the deformation is stored within the deformed or strained body as *strain energy*. When the body behaves elastically, as in a spring, the strain energy will be released when the object returns to its original dimensions.

Thus a spring is merely a device for storing energy; for example, the main spring in a watch stores the energy input as it is wound and gradually expends it in 'driving' the watch.

7.6 POTENTIAL ENERGY

Consider a mass of m kg which is lifted through a vertical height of h m against the gravitational resistance of mg N.

$$\begin{aligned}\text{Work done} &= \text{force} \times \text{distance} \\ &= mg \times h \text{ N m} \\ &= mgh \text{ J}\end{aligned}$$

This work done will be stored as potential energy while the object remains at this height. If the object is allowed to fall from this height then the potential energy will obviously decrease and, as we will see later (section 7.8), the loss in potential energy will equal the gain in kinetic energy due to the velocity of the fall.

This principle is used in pile-driving techniques on construction sites, and in hydro-electric generation schemes the potential energy possessed by the water stored in reservoirs at high levels is converted to kinetic energy to drive the turbines to generate electricity. By this method energy can be obtained without any depletion of the world's fossil fuels. Unfortunately in the United Kingdom geographical features limit the supply of this form of energy; furthermore suitable sites are situated away from the urban and industrial centres for which the energy is required.

Fundamentally any body will possess potential energy whenever work is done on it against gravitational attraction, that is, when it is moved away from the centre of the Earth. Instead of constantly referring back to the centre of the Earth, it is usual to specify a datum when referring to potential energy; often we are only concerned with changes in potential energy occurring as a result of a change in height.

7.7 KINETIC ENERGY

Consider a mass of m kg at rest being subjected to a constant force, then following Newton's second law ($F = ma$) an acceleration will be produced causing the body to start to move. Now

$$\text{work done} = \text{force} \times \text{distance}$$
$$= ma \times s \qquad (7.1)$$

Since the force remains constant the acceleration will be uniform and from section 5.1 we have

$$v^2 = u^2 + 2as$$

but $u = 0$ therefore

$$v^2 = 2as$$
$$a = \frac{v^2}{2s}$$

Substituting this into equation 7.1

$$\text{work done} = m \times \frac{v^2}{2s} \times s$$
$$= \frac{mv^2}{2}\ \text{J}$$

An alternative derivation of this formula is illustrated in figure 7.4. A force F acting on a stationary mass m acting through a distance s produces a velocity v. Substituting the values obtained into the standard work-done formula

$$\text{work done} = \text{force} \times \text{distance}$$
$$= ma \times \frac{vt}{2}$$
$$= m\left(\frac{v}{t}\right) \times \frac{vt}{2}$$
$$= \frac{1}{2}mv^2$$

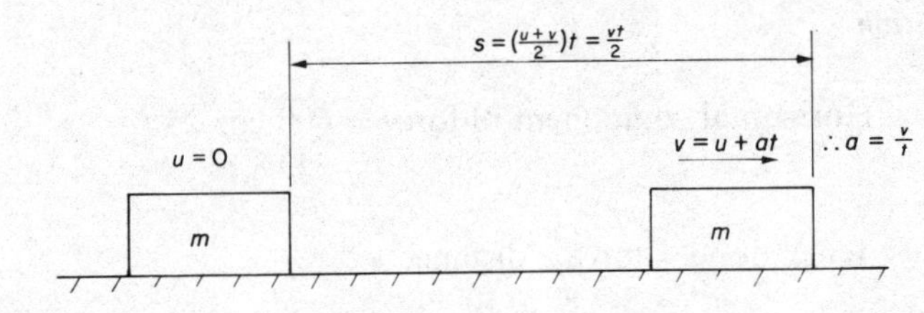

Figure 7.4

This work done is stored within the body while it is moving at velocity v and is known as the kinetic energy. To ensure correct units, the mass must be measured in kg and the velocity in m/s, since

$$\text{kinetic energy} = \frac{1}{2}mv^2$$

m = force/acceleration = N/(m/s^2), therefore

$$\text{kinetic energy} = \frac{1\,\text{N}}{2\,\text{m}}\text{s}^2\left(\frac{\text{m}^2}{\text{s}^2}\right) = \text{N m}$$

This energy can be used by the moving body not under power to overcome any resistance to motion.

$$\text{Kinetic energy} = \text{resistance} \times \begin{array}{l}\text{distance travelled in}\\ \text{coming to rest}\end{array}$$

Up to this point in our studies we have dealt with the energy possessed by a body as a result of its linear motion. However, a body that is rotating will also possess kinetic energy due to its rotation, for example, a flywheel is designed to store this energy effectively.

A moving vehicle such as a car will possess linear kinetic energy (or kinetic energy of translation) and its rotating parts such as wheel-and-axle assemblies will also possess kinetic energy of rotation. The total kinetic energy possessed will be the sum of these.

Example 7.7

Determine the work done in causing a mass of 350 kg to move at 5 km/h from rest.

Solution

$$5 \text{ km/h} = \frac{5 \times 1000}{60 \times 60} = 1.39 \text{ m/s}$$

$$\text{Kinetic energy} = \tfrac{1}{2} mv^2$$
$$= \frac{350\,(1.39)^2}{2} = 338.1 \text{ J}$$

Example 7.8

A car of mass 2000 kg is travelling at 50 km/h when the engine is switched off. Calculate the kinetic energy possessed and determine the distance the vehicle will travel against a resistance of 300 N.
Solution

$$\text{Kinetic energy} = \tfrac{1}{2} mv^2$$
$$= \frac{2000}{2}(13.9)^2 \text{ J}$$

$$= 193.2 \text{ kJ}$$
$$193.2 \times 10^3 = \text{force} \times \text{distance}$$
$$\text{distance} = \frac{193.2 \times 10^3}{300} = 644 \text{ m}$$

7.8 CONSERVATION OF ENERGY

The law of conservation of energy states that the total energy in a system remains constant unless the system is acted upon by an external force. We have already seen that energy cannot be created or destroyed but only transformed from one form to another, and the law of conservation of energy is an extension of this statement. The total content of energy in the universe will always remain the same; Einstein has shown that the mass of a body is a measure of the quantity of energy contained in it and it has been shown that it is possible (in a nuclear reactor) for a slight loss in mass to result in a large gain in energy. This is why nuclear weapons are so destructive. The deterrent effect of nuclear warfare may have prevented a major outbreak of hostilities between the world powers. However, the full implementation of plans for the peaceful use of atomic power has been handicapped by problems of control of reactions involving the splitting of the nucleus of such fuels as uranium and plutonium and the disposal of the radio-active waste materials.

In the light of Einstein's work on relativity it is more correct to say that the sum total of the mass and energy in the universe should be regarded as constant.

We have been discussing the various forms of mechanical energy and there are many practical applications whereby a conversion from one form of mechanical energy to another takes place to which we can apply the law of conservation of energy. We have already mentioned the spring as a means of storing energy, but a conversion of energy also takes place—the strain energy released may be converted to the kinetic energy possessed by the moving parts of the watch.

Other machines may be designed to arrange more elaborate transformations but one simple transformation which has been referred to already is that between potential and kinetic energy. For example, in pile-driving a large mass m is allowed to fall freely through a height h losing its potential energy and gaining kinetic energy; thus

$$mgh = \tfrac{1}{2}mv^2$$

Hence the velocity after a fall through height h may be obtained, where

$$v = \sqrt{(2gh)}$$

Note that this velocity is independent of the mass, thus confirming our earlier observations (section 5.3) that, *neglecting resistances*, all masses will fall with equal velocity.

An object losing height in a non-vertical direction, for example, an aircraft coming in to land, will also be losing potential energy. In the case of a vehicle or an object rolling down a slope, its loss of potential energy may be used to overcome the resistances to motion, with any remaining energy being the kinetic energy due to motion.

Example 7.9

A pile-driver of mass 700 kg falls freely through a height of 2 m on to a pile of mass 150 kg. Determine the velocity with which the driver hits the pile. If at impact 2.5 kJ of energy are lost due to heat, sound, etc., the remaining energy being possessed by the pile and driver as they are driven together into the ground a distance of 150 mm, find the common velocity after impact and the average resistance of the ground.

Solution Potential energy of pile-driver is converted into kinetic energy, thus

$$\begin{aligned} \text{potential energy} &= \text{kinetic energy} \\ mgh &= \tfrac{1}{2}mv^2 \\ v &= \sqrt{(2gh)} \\ &= \sqrt{(2 \times 9.81 \times 2)} \\ &= 6.264 \text{ m/s} \end{aligned}$$

the velocity with which the pile driver hits the pile. Kinetic energy of pile-driver before impact is given by

$$\begin{aligned} \text{kinetic energy} &= \tfrac{1}{2}mv^2 \\ &= \tfrac{1}{2}(700)(6.264)^2 \\ &= 13.733 \text{ kJ} \end{aligned}$$

$$\begin{aligned} \text{Kinetic energy after impact} &= 13.733 - 2.5 \\ &= 11.233 \text{ kJ} \end{aligned}$$

This is possessed by pile-driver and pile as they move together (total mass 850 kg).

$$\begin{aligned} \text{Kinetic energy} &= \tfrac{1}{2}mv^2 \\ 11.233 \times 10^3 &= \tfrac{1}{2} \times 850 \times v^2 \\ 26.43 &= v^2 \\ v &= 5.14 \text{ m/s} \end{aligned}$$

which is the common velocity after impact. The kinetic energy after impact is absorbed in overcoming resistance of ground for distance of 150 mm, where

$$\begin{aligned} \text{kinetic energy} &= \text{resistance} \times \text{distance} \\ 11.233 \times 10^3 &= \text{resistance} \times 0.150 \end{aligned}$$

therefore

$$\begin{aligned} \text{resistance} &= \frac{11.233 \times 10^3}{0.150} \\ &= 74\,886 \text{ N} \end{aligned}$$

$$\text{average resistance} = 74.886 \text{ kN}$$

Example 7.10

A car of mass 500 kg is uniformly accelerated from rest up an incline of 1 in 30, covering a distance of 100 m in reaching a velocity of 70 km/h. Determine the work done against gravitational effects, the gain in kinetic energy, the total work done and the total force required.

Solution This problem can be solved using the concept of potential or kinetic energy and by the application of dynamic theory.

By the energy method

$$\begin{aligned} \text{gain in potential energy (against gravity)} &= mgh \\ &= 500 \times 9.81 \times 3.33 \text{ J} \\ &= 16.33 \text{ kJ} \end{aligned}$$

$$\begin{aligned} \text{gain in kinetic energy} &= \tfrac{1}{2}mv^2 \\ &= \frac{500}{2}(19.44)^2 \text{ J} \\ &= 94.48 \text{ kJ} \end{aligned}$$

$$\begin{aligned} \text{total gain in energy} &= 16.33 + 94.48 \\ &= 110.8 \text{ kJ} \end{aligned}$$

$$\text{Energy} = \text{force} \times \text{distance}$$

$$\text{force} = \frac{110.8}{100} = 1.1 \text{ kN}$$

By the dynamic method

$$\text{gravitational force} = mg = \frac{500 \times 9.81}{30} = 163.5 \text{ N}$$

To find acceleration

$$v^2 - u^2 = 2as$$
$$a = \frac{v^2}{2s} = \frac{(19.44)^2}{2 \times 100} = 1.889 \text{ m/s}^2$$

$$\text{Accelerating force} = 500 \times a = 500 \times 1.889 = 944.5 \text{ N}$$

Work done = force × distance
Work done against gravity = 163.5 × 100 J = 16.35 kJ

Work done by accelerating force = 944.5 × 100 J = 94.45 kJ

Total work done = 110.8 kJ
Total force applied = 163.5 + 944.5 N = 1.108 kN

Example 7.11

The hammer of an Izod-type impact testing machine has a mass of 7 kg and can be taken as being concentrated at a radius of 1 m. It is raised to a horizontal position and then allowed to swing freely in a vertical circle on a pivot imagined to be frictionless. Calculate the velocity of the hammer and its kinetic energy at its lowest position. Assuming that at this point it fractures a specimen which absorbs 25 J of energy, determine the height to which the hammer will swing and its angle of inclination to the vertical.

Solution Potential energy = kinetic energy, therefore

$$mgh = \tfrac{1}{2} mv^2$$
$$v = \sqrt{(2gh)} = \sqrt{(2 \times 9.81 \times 1)} = 4.43 \text{ m/s}$$

$$\text{kinetic energy} = \tfrac{1}{2} mv^2 = \tfrac{1}{2} \times 7 \times (4.43)^2 = 68.67 \text{ J}$$

(Alternatively, potential energy = $mgh = 7 \times 9.81 \times 1 = 68.67$ J).

After impact the hammer will have (68.67 − 25) J of kinetic energy which will be converted into potential energy.

$$\text{Potential energy} = mgh$$
$$43.67 = mgh$$

therefore

$$h = \frac{43.67}{7 \times 9.81} = 0.636 \text{ m}$$

that is, 0.364 m below its original height.

Angle of inclination = 68.65°

7.9 POWER—THE RATE OF DOING WORK

So far in our discussions of energy being converted or expended to do work we have not considered the effect of the time taken to convert the energy or to do the work. The rate at which the energy is converted or the rate at which the work is done is obviously a very important consideration. It will require a much stronger or more powerful man to dig an allotment more quickly than a weaker man and if the man is equipped with more power—machine power in the form of a rotovator or mechanical digger—the work will be done even more quickly.

When a particular job must be carried out more quickly, more power must be supplied, either in the form of more men or more powerful equipment.

Power, both in the technical and everyday sense, is the term used to refer to the rate at which work is done, or is capable of being done; thus power is the work done divided by the time taken.

$$\begin{aligned}\text{Power} &= \text{rate of work done}\\ &= \text{rate at which energy is expended}\\ &= \frac{\text{work done}}{\text{time}}\\ &= \frac{\text{force} \times \text{distance}}{\text{time}} \text{ or force} \times \frac{\text{distance}}{\text{time}}\\ &= \text{force} \times \text{velocity}\end{aligned}$$

This gives the units of

$$\text{power} = \frac{\text{force (N)} \times \text{distance (m)}}{\text{time (s)}}$$

$$= \text{N m per s} = \text{J per s} = \text{W}$$

1 W is the rate of working equal to 1 J per s or 1 N m per s.

The use of SI units gives the very real advantage that the same unit is used for power irrespective of the form of energy or work done being considered. Until the adoption of this system of units in the United Kingdom the traditional unit of mechanical power was the horsepower, attributed to James Watt.[2] One horsepower was an inaccurate estimate of the power possessed by a horse (said to be capable of 33000 ft lb per minute) used by Watt to compare the capabilities of his engines to the capabilities of the horses they were replacing.

1 James P. Joule (1818–89) of Salford was a pupil of Dalton. His name is commemorated by the mechanical equivalent of heat in the imperial system of units; students, technicians and engineers had to remember that 778 ft lb of mechanical energy were equivalent to one British thermal heat unit.

Joule followed up earlier work by Benjamin Rumford, who tried to measure the heat produced during the boring of a gun barrel by using it to heat a known amount of water. Joule arranged a closed container with a specified amount of water which was capable of being agitated by a series of paddles driven by falling weights. As the weights drove the paddles agitating the water the temperature of the water rose and was measured; Joule found that it required the expenditure of 772 ft lb of mechanical energy to raise the temperature of 1 lb of water through 1°F. This amount of heat was known as the British thermal unit of heat. The standard equivalent was later modified to the universally accepted value of 778 ft lb to 1 Btu.

Joule later collaborated with Kelvin on work dealing with conservation of energy and on temperature scales.

Today his name is used for the unit of energy or work done.

2 James Watt (1736–1819) showed some mathematical ability at school and had the opportunity to acquire practical skills at an early age. He was apprenticed to a London instrument maker (at a fee of 20 guineas a year).

He returned to Glasgow in 1756 and through his friendships with Professors Black and Anderson was allowed to set up a workshop as instrument maker to the University.

In 1764 he was asked to improve a scale model of an engine developed in 1712 by Thomas Newcomen to pump water out of mines in Cornwall. Newcomen himself had improved an earlier idea by Denis Papin whereby a stroke in one direction was produced by the steam from boiling water, the reverse stroke being produced as the result of a partial vacuum inside the cylinder, produced in turn by condensing the steam, and the atmospheric pressure. The heating and condensation were carried out by heating and cooling the cylinder, which Watt realised was wasteful so he piped the steam to a separate vessel to be condensed. This enabled the engine to work much faster and produce such substantial savings in fuel that Britain entered the steam age, which was to revolutionise transport and industry.

Watt applied for a patent for his engine in 1768 and later in 1775 an Act of Parliament was passed to extend the period of his patent; he also went into partnership with Matthew Boulton to exploit the application of his invention. The operation of the engine was improved in 1780 by making the engine 'double-acting'. During this time Watt worked as a surveyor and also invented a form of printing press.

The use of a crank, and also of a 'sun and planet' mechanism to convert the linear motion to rotary motion ushered in the age of steam locomotion (with the work of Richard Trevithick and George Stephenson) and the Industrial Revolution. There were benefits to society but the industrialisation process also involved child and female labour, which necessitated the passing, in 1803, of the forerunner of the current Factories Act and the Health ano Safety at Work Act.

TO THE STUDENT

At the end of this chapter you should be able to

(1) define energy and recognise that it can exist in different forms
(2) recognise that the unit of work is the joule, and that this unit may be used to measure all forms of energy
(3) calculate the work done by both constant and variable forces
(4) define kinetic and potential energy and solve simple problems
(5) state the law of conservation of energy and apply this law to typical cases of energy conversion
(6) define power as the rate of doing work and solve simple problems using the watt
(7) complete exercises 7.1 to 7.15.

EXERCISES

7.1 A machine tool tablc has a mass of 160 kg and carries a vice component which weighs 110 N. Determine the work done if the table assembly is lifted through a height of 80 mm.

7.2 A shaping machine makes 40 cutting strokes per minute. If the length of each stroke is 450 mm and the cutting force on the tool is 500 N, determine the work done per minute.

7.3 The cutting force on a lathe tool is 150 N and the work being turned is 70 mm in diameter, revolving at 120 rev/min. Determine the work done per minute.

7.4 The pressure in a cylinder 80 mm in diameter varies linearly from 140 kN/m^2 at the beginning of the stroke to 10.5 kN/m^2 at the end. If the stroke is 150 mm long, determine the work done per stroke.

7.5 A load is gradually applied to a helical spring and when the extension is 40 mm it is found that the load is 60 N. Determine the work done in producing this extension.

7.6 A buffer spring whose stiffness is 15 kN/m is compressed a distance of 20 mm. Determine the work done.

7.7 During a machining operation the force on the cutting tool was found to vary as follows.

Distance (m)	0	0.75	1.5	2.25	3.00	3.75	4.50
Force (N)	500	450	430	400	420	430	400

Plot a force–distance graph and from it determine (a) the work done per stroke and (b) the average cutting force.

7.8 During a broaching operation the cutting force varies during the stroke as follows.

Length of stroke (mm)	0	25	50	75	100	125	150
Force (N)	80	75	66	60	53	48	45

From a force–distance graph determine (a) the work done per stroke and (b) the average cutting force.

7.9 A casting is pulled a distance of 6 m along a horizontal surface by a rope inclined at 30° to the horizontal. If the tension in the rope is 275 N determine the work done.

7.10 A body of mass 70 kg is pushed along a horizontal surface a distance of 10 mm. If the coefficient of friction between the surface and the mass is 0.3, determine the work done and the total force necessary if the force is applied at an inclination of 20° to the direction of motion.

7.11 A body of mass 20 kg is allowed to fall freely through a height of 175 m. Determine its final velocity.

7.12 A lorry of mass 2 t (2000 kg) travelling along a level road at 40 km/h is brought to rest in a distance of 18 m. Determine its kinetic energy while travelling at this speed and the average braking force applied.

7.13 A mass of 2.5 t falls through at vertical height of 5 m striking the ground and penetrating a distance of 250 mm before it is brought to rest. Determine the velocity with which it hits the ground, and the average resistance of the ground.

7.14 A bullet of mass 14 g travelling horizontally at 450 m/s penetrates a fixed block of wood 250 mm thick. If it experiences an average resistance of 200 N determine the velocity with which the bullet emerges from the block.

7.15 A car of mass 300 kg travelling at 50 km/h is brought to rest in a distance of 20 m. Determine the average braking force applied and the time taken.

NUMERICAL SOLUTIONS

7.1 134.4 J
7.2 9000 J per min
7.3 66 J per min
7.4 225 J
7.5 1.2 J
7.6 300 J
7.7 1949 J; 433 N
7.8 9.13 J; 61 N
7.9 2.46 N m
7.10 197.6 J; 2060 N
7.11 58.6 m/s
7.12 123 kJ; 6860 N
7.13 9.9 m/s; 49 kN
7.14 320 m/s
7.15 1428 N; 65.7 s

8 Heat and its Applications

The object of this chapter is to give the student an understanding of the nature of heat as a form of energy and to enable him to recognise the effects of heat as encountered in an engineering context.

8.1 HEAT AND ENERGY

Heat is a form of energy so we must first revise what is meant by the term energy. One definition of energy is 'the ability to do work' (section 7.2). If heat is energy, it must be measured in energy units, which are joules (see p. 84). Recognising that heat is a form of energy we can look at some of the effects it is likely to produce.

8.2 EFFECTS OF HEAT

The most easily recognised effect of heat is its *ability to make something hotter*, which can be detected by our sense of touch and by other more accurate methods as well. The second effect is the ability of heat to *change the state* of a substance, by which we mean melting and evaporation. A further effect that we can detect is the ability of heat to *change the size* of a substance. This is less easy to detect with the senses but simple instruments can make it easily discernible. Other effects include the ability of heat to *generate small electric currents* in certain conductors and of course the *production of visible light* from an intensely heated substance (known as incandescence).

8.3 SOURCES OF HEAT

Heat is obtained by conversion of other forms of energy, a process which is usually fairly simple and spontaneous, that is, a process which, once started, proceeds on its own. Table 8.1 shows some of the methods.

Table 8.1

Source	*Original Energy Form*	*Conversion Appliance*
Electricity supply	Electrical	Heating element (e.g. fires, kettles, soldering irons
Fuel (e.g. oil, gas, coal)	Chemical	Combustion chambers (furnaces, engines)
Mechanical movement	Work	Friction (in bearings, brakes, friction welding)
Nuclear reactions	Nuclear	Nuclear reactors in power stations

As these examples show, heat is often generated for useful purposes, but on some occasions it is a nuisance and wastes energy because it can leak away very easily.

We should note here that energy *cannot be created or destroyed* by ordinary means, so heat must *always* be obtained from another form of energy.

8.4 HEAT TRANSFER

Heat energy is able to pass from one location to another by three quite distinct methods. Sometimes it may be moved or transmitted by all three at once, sometimes by only one or two methods. The methods are as follows.

8.4.1 Conduction

Let us consider how we might observe this in a solid material such as a metal bar. If one end of the bar is in contact with a flame or other heat source it rapidly becomes hot while the other end remains cool. After some time, however, we can detect that the cool end has become warm or even hot, although not as hot as the heated end.

To explain this *flow* of heat along the bar we assume that the metal is composed of extremely small particles called molecules, which are in constant motion. In a solid this molecular motion is in the form of imperceptible *vibration* about a fixed point, which explains how the solid manages to keep its shape. As the metal is heated, the energy absorbed is in the form of energy of motion of the molecules so they have to vibrate more rapidly. This results in collisions with neighbouring molecules which absorb some energy from the collision and vibrate faster. This goes on down the line of neighbours until energy has reached the other end of the bar. Note though, that *no molecules* have moved down the bar (see figure 8.1). The transfer movement resembles waves of motion which pass along a string that is vibrated from one end.

Heat can be conducted through liquids and gases by the same process but usually with more difficulty. The increase in the speed of movement of the molecules when heated means that their kinetic (movement) energy level is increased. We call this the *internal energy* of a substance and we recognise its level by observing the temperature of the substance. Higher temperature means higher internal energy level. This in turn means more violent movement of the molecules which enables them to pass on heat energy to their neighbours more easily. This is easily confirmed by experiments.

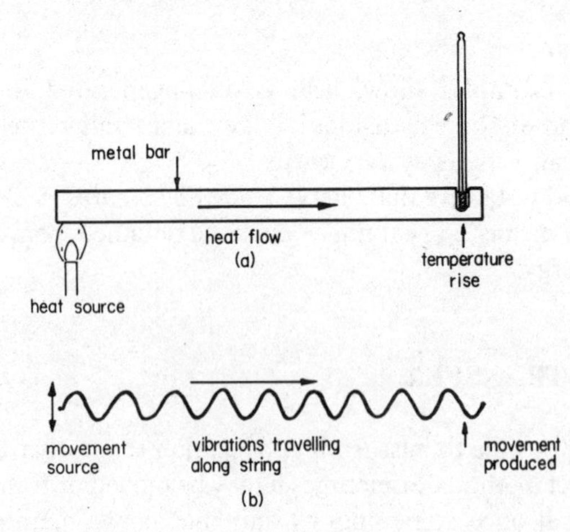

Figure 8.1

8.4.2 Convection

In this process heat is absorbed by a substance which is then *moved* from one location to another location, there to *give up* its heat to a suitable receiver. A very good example of this is a fan heater, where a stream of air is used to absorb the heat from a hot element and then deliver the heat to other parts of a room. Sometimes the air movement takes place without the use of a fan (natural convection)—see figure 8.2. This effect is produced by the reduction in density of heated air, enabling it to be displaced upwards by colder air from other sources. Note that the cold air displaces the hot air; the hot air will *not* rise on its own. The cold air then becomes heated in turn and is displaced by more cold air, producing a continuous process (called a convection current).

Convection can only occur with materials which are *fluids* (that is, capable of flowing) such as liquids and gases.

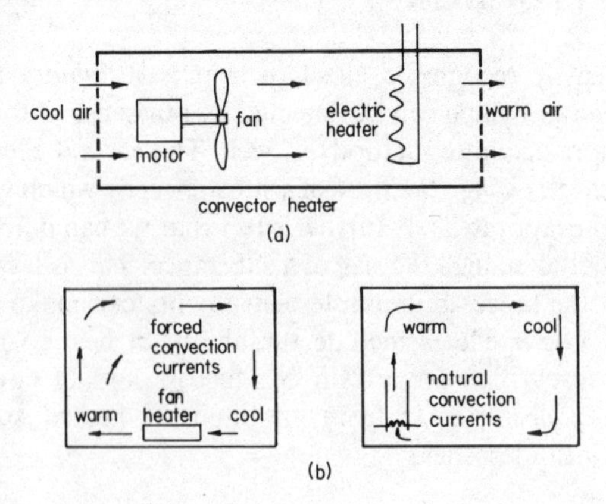

Figure 8.2

8.4.3 Radiation

Energy may be emitted by a third process known as thermal radiation, which is quite distinct in nature from the other two. This involves the production of energy waves known in general as *electromagnetic radiation,* which includes radio waves, thermal radiation, light waves, X-rays and γ-rays produced from nuclear sources. Thermal radiation is emitted by all bodies and requires no *medium* for transmission, that is, it will cross a vacuum if necessary. Many materials will transmit this radiation, while others absorb it completely. The result of absorbing thermal radiation is to raise the temperature of the absorbing material, so we recognise this as a method of *heat transfer.* Thermal radiation has a wavelength slightly greater than that of the visible red light and is often referred to as *infrared* radiation. In many ways it behaves like light itself and can be reflected and focused by mirrors and lenses (figure 8.3). (An electric fire makes use of this with a silvered reflector to direct the heat forwards into the room.) In the same way dark and dull surfaces absorb radiant heat more readily than light shiny surfaces. This also applies when we examine the emission of radiant heat; dark dull surfaces emit more readily than shiny surfaces. (Note that the term 'radiator' is misused when describing the hot-water room heater; such devices transfer their heat mainly by *convection* to the surrounding air.)

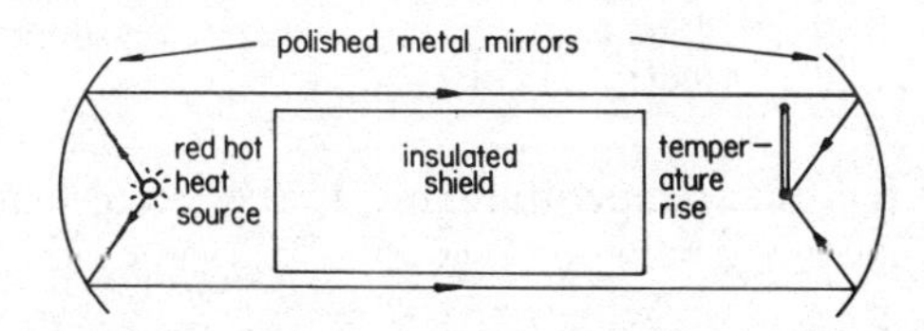

Figure 8.3 Demonstration that radiant heat is similar to light

8.5 TEMPERATURE

We have already made reference to the *temperature* of a substance and recognise it as a means of measuring the energy 'level' or 'intensity' in a body. The expression 'potential' is often used (as in electricity) since temperature is the effect which produces *heat flow.* We should recognise that temperature does *not* directly measure the heat energy content of a body—rather we might describe it as a 'degree of hotness'. It is very useful to measure temperature level by comparing the effect produced on a thermometer; if two bodies are able to produce the same effect we say that they must have the same temperature.

8.5.1 Temperature Scales

For convenience we allocate numerical values to the response of the thermometer, using *temperature scales.* To create a temperature scale we need two fixed points such as the temperatures of melting ice and boiling water, because they are easy to keep at a steady value. The space between these two positions on a thermometer is then divided into a convenient number of intervals called *degrees.* Figure 8.4 shows how this is done for the Celsius and Fahrenheit scales.

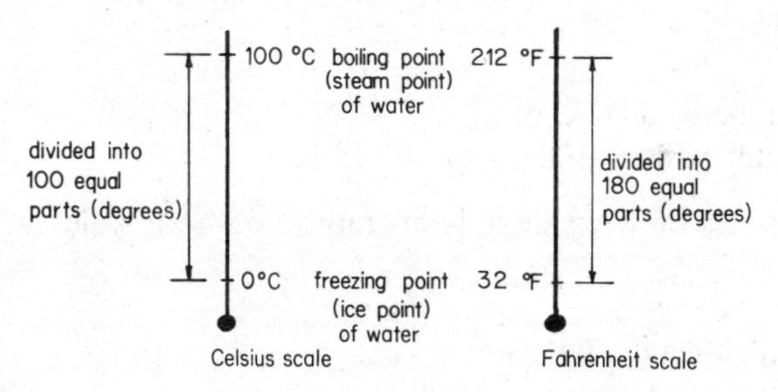

Figure 8.4

The *Celsius*[1] scale is the more commonly used in scientific work and in its extended form has been adapted as the *Kelvin*[2] temperature scale. It is possible to cool down a body far below 0°C and this has led to the idea of an absolute zero temperature which cannot be reduced any further. By using a special instrument called a *gas thermometer* it is possible to establish this value as −273.15°C which is made the lower fixed point of the Kelvin scale. At this temperature the molecular vibration referred to in section 8.4.1 is assumed to disappear. The upper fixed point is established by using

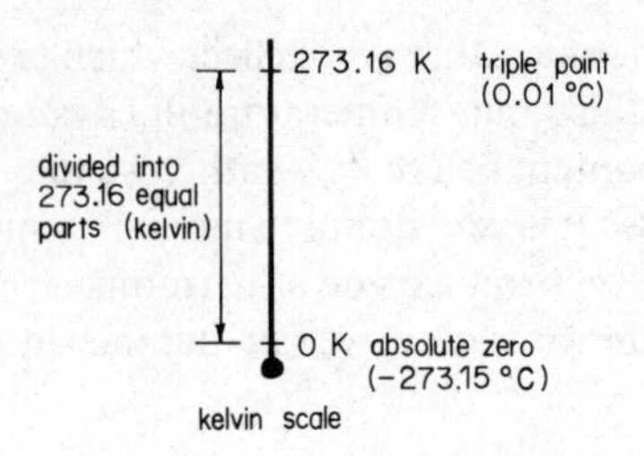

Figure 8.5

a special vessel called a *triple point cell* which keeps water, ice and water vapour in equilibrium with each other at a very reliable steady temperature of 0.01° C. So we have a new and reliable scale of temperature as shown in figure 8.5. Note that the new degrees are called kelvins *not* degrees kelvin, symbol K. The scale is immensely useful in the study of heat in engineering and scientific work. For convenience the Celsius scale is used for general measurement and its degrees are the same size as kelvins. For example

ice melts at 0 °C or 273.15 K
water boils at 100 °C or 373.15 K

The *interval* between these temperatures on both scales is 100 K

8.6 CALORIMETRY

This is the name given to methods of calculating the quantity of heat absorbed or emitted during a technical process.

8.6.1 Specific Heat Capacity

Different materials absorb different amounts of heat when their temperature is raised by a given amount (see figure 8.6). The materials are said to have different *specific heat capacities.* The specific heat capacity of a substance is defined as the quantity of heat (J) required to produce *unit temperature change* (1 K) in *unit mass* (1 kg) of the substance. The symbol used is *c* and the units of

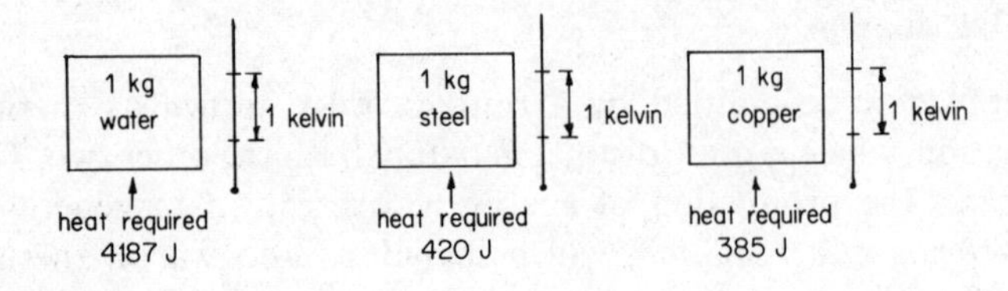

Figure 8.6

measurement are J/kg K (or J kg^{-1} K^{-1}). For example, the specific heat capacity of steel is 420 J/kg K, which means that 420 J of heat must be supplied to 1 kg of steel to raise its temperature by 1 K. Other typical values are

Copper 385 J/kg K (or 0.39 kJ/kg K to two decimal places)
Water 4187 J/kg K (or 4.19 kJ/kg K to two decimal places)

8.6.2 Calculation of Heat Lost or Gained

When the temperature of a body is changed we can calculate the quantity of heat involved if we know the *mass*, the *specific heat capacity* and the *temperature change.*

$$\text{Heat flow, } Q = \text{mass} \times \begin{matrix}\text{specific heat}\\ \text{capacity}\end{matrix} \times \begin{matrix}\text{temperature}\\ \text{change}\end{matrix}$$

$$= mc(t_2 - t_1)\,\text{J}$$

where t_2 is the final temperature and t_1 the initial temperature.

Example 8.1

Calculate the heat absorbed by a car radiator made of copper if its mass is 2 kg and the temperature is increased from 10 °C to 45 °C.
Solution

$$Q = 2 \times 385 \times (45 - 10)$$
$$= 26950\ \text{J} = 26.95\ \text{kJ}$$

8.6.3 Heat Capacity of a Body

Heat capacity (or thermal capacity) is defined as the quantity of heat required to produce unit temperature change *of the body*.

Consider the car radiator described in example 8.1. If $(t_2 - t_1) = 1$ K, then

$$Q = 2 \times 385 = 770 \text{ J/K}$$

that is

$$\text{heat capacity} = \text{mass} \times \text{specific heat capacity}$$

8.6.4 Heat Capacity of a Composite Body

Where a body is made of several materials its *total* heat capacity is the *sum* of the separate heat capacities of the different masses of each material. Thus

$$\text{total heat capacity} = m_1 c_1 + m_2 c_2 + m_3 c_3 + \dots$$

8.6.5 Water Equivalent

For convenience we often express the heat capacity of a body in terms of an *equivalent* mass of water (*not* an *equal* mass). Thus we can say that the water equivalent of a body is that mass of water which has the *same heat capacity* as the body.

Let m_b = mass of body, c_b = specific heat capacity of body, m_w = water equivalent, c_w = specific heat capacity of water. Then

$$m_w c_w = m_b c_b$$

therefore

$$m_w = m_b \frac{c_b}{c_w}$$

Example 8.2

Calculate the water equivalent of the copper radiator described in example 8.1.

Solution

$$m_w = 2 \times \frac{385}{4187} = 0.184 \text{ kg}$$

The advantage of using water equivalents is that several heat capacities can be added together without having to consider the specific heat of each material.

8.6.6 Mixtures

When two bodies at different temperatures are brought into contact, heat will flow *from the hotter* body *to the cooler* body until equilibrium is reached, that is, both bodies have the same temperature (see figure 8.7). Such a process is called *mixing* and the combined bodies are a *mixture.* We may assume that

$$\begin{matrix}\text{heat lost in} \\ \text{cooling hot body}\end{matrix} = \begin{matrix}\text{heat gained in} \\ \text{warming cold body}\end{matrix}$$

Let t_1 = temperature of hot body, A, t_2 = temperature of cold body, B, and t_3 = final temperature of both bodies. Then

$$m_A c_A (t_1 - t_3) = m_B c_B (t_3 - t_2)$$

We can use this equation to find certain unknown quantities.

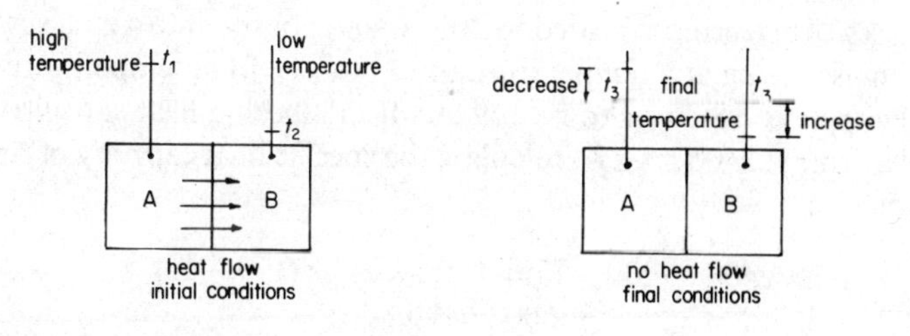

Figure 8.7 Thermal mixing process

Example 8.3

A chisel of mass 0.5 kg and specific heat capacity 420 J/kg K is

heated to 450 °C. It is then quenched in 5 kg of oil of specific heat capacity 2000 J/kg K at a temperature of 15 °C. Calculate the final temperature of the mixture.
Solution Using the mixture equation

$$0.5 \times 420\,(450 - t_3) = 5 \times 2000\,(t_3 - 15)$$
$$94500 - 210t_3 = 10000t_3 - 150000$$
$$10210t_3 = 244500$$
$$t_3 = \frac{244500}{10210} = 23.95°\text{C}$$

Example 8.4

A metal block at 15 °C is lowered into a drum containing 110 kg of water at 100 °C, whereupon the temperature falls to 73 °C. If the specific heat capacity of the metal is 913 J/kg K calculate the mass of the block.
Solution

$$m_A \times 913\,(73 - 15) = 110 \times 4187\,(100 - 73)$$
$$m_A = \frac{110 \times 4187 \times 27}{913 \times 58}$$
$$= 234.8 \text{ kg}$$

Example 8.5

1.5 kg of glycerine is heated to 200 °C and poured into a zinc vessel of mass 1.2 kg at a temperature of 15 °C. The final temperature of glycerine is observed to be 179 °C. If the specific heat capacity of glycerine is 2400 J/kg K, calculate the specific heat capacity of zinc.
Solution

$$1.5 \times 2400 \times (200 - 179) = 1.2 \times c_{Zn} \times (179 - 15)$$
$$75600 = 196.8c_{Zn}$$
$$c_{Zn} = \frac{75600}{196.8}$$
$$= 384.1 \text{ J/kg K}$$

8.7 CHANGE OF STATE

It has been found by experimental observation that changing from solid to liquid and liquid to gas requires a supply of heat which has a quite different effect from the processes already described.

8.7.1 Latent Heat

If a graph such as figure 8.8 is plotted of heat added to a *fixed amount* of water (for example, 1 kg), starting with ice and proceeding through to steam we can see that in each *phase* the addition of heat produces a *rise* in temperature. This can be detected with our senses and is often called *sensible* heat input. On the other hand we see that during the transition from one phase to another a quantity of heat has to be added which does *not* alter the temperature. It is said to be 'hidden' or *latent* heat. The reason for this extra energy, needed for a *change of state*, is connected with the molecules and their effect on neighbouring molecules. In a solid, the vibrating molecules remain in more or less the same position because they tend to cling together with a force of attraction. When melting occurs, extra energy is needed to overcome some of this attraction and give the molecules a *greater degree of freedom* so they can move about within the liquid. When the reverse process occurs, and a solid is formed, the extra energy is released by the substance while the temperature stays constant. This energy quantity is called the *latent heat of fusion.*

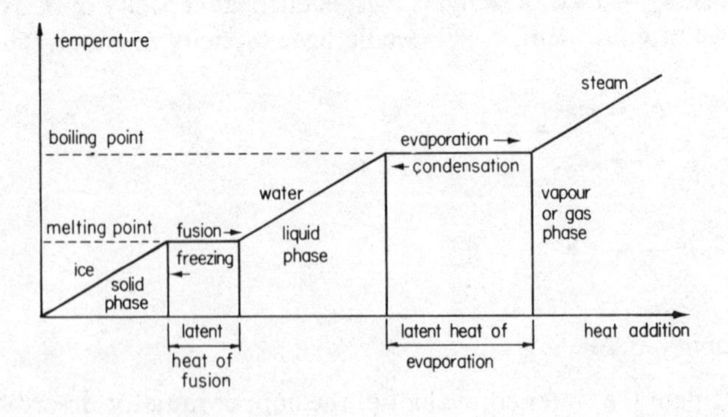

Figure 8.8

A similar process occurs when liquids evaporate to form gases. The molecules in the liquid are still close enough to exert some attraction on each other, but during evaporation they absorb a *further* quantity of energy to break completely free from their neighbours. In gases the molecules can roam about independently and have the greatest degree of freedom. The energy needed to break free is the *latent heat of vaporisation.* (Elsewhere this may be referred to as *specific enthalpy of evaporation*, which means the same thing.) Under normal atmospheric pressure 1 kg of water requires 2257000 J of latent heat to convert it to 1 kg of steam, *at the same temperature* (or 2257 kJ/kg). At normal atmospheric pressure 1 kg of ice needs 335000 joules to convert it to 1 kg of water, *at the same temperature* (or 335 kJ/kg). Other substances need different amounts.

Example 8.6

Calculate the heat required to convert 15 kg of water to steam at the same temperature.
Solution

$$\text{Latent heat required} = 15 \times 2257 = 33855 \text{ kJ}$$

Example 8.7

How much heat must be extracted by a refrigerator if it makes 250 ice cubes each at 0°C having a mass of 50 g?
Solution

$$\text{Latent heat removed} = 250 \times \frac{50}{1000} \times 335 = 4187.5 \text{ kJ}$$

Example 8.8

Calculate the heat removed in condensing 200 kg of ethyl alcohol vapour in a distillery if the latent heat of vaporisation of alcohol is 850 kJ/kg.

Solution

$$\text{Latent heat removed} = 200 \times 850 = 170000 \text{ kJ}$$

8.7.2 Latent Heat in Mixtures

When a solid such as ice is mixed with a liquid such as water, heat flows from the water to the ice and melts it, then raises its temperature until equilibrium is reached.

$$\begin{matrix}\text{Heat lost by} \\ \text{cooling water}\end{matrix} = \begin{matrix}\text{heat gained by} \\ \text{melting ice}\end{matrix} + \begin{matrix}\text{heat gained by} \\ \text{melted ice}\end{matrix} \qquad (8.1)$$

Example 8.9

0.1 kg of ice at 0°C is mixed with 1.5 kg of water at 35°C. Calculate the final temperature of the mixture.
Solution Using equation 8.1 (working in kJ)

$$\underset{\text{(sensible heat)}}{1.5 \times 4.187(35 - t_3)} = \underset{\text{(latent heat)}}{0.1 \times 335} + \underset{\text{(sensible heat)}}{0.1 \times 4.187(t_3 - 0)}$$

$$219.8 - 6.28t_3 = 33.5 + 0.4187t_3 - 0$$

$$186.3 = 6.699t_3$$

$$t_3 = \frac{186.3}{6.699} = 27.8°\text{C}$$

Example 8.10

An industrial process uses steam injection to heat a treatment vat containing 100 kg of water. If the water is initially at 15°C and is heated to 70°C, how much steam at 100°C is used?
Solution

$$\begin{matrix}\text{Latent heat lost} \\ \text{by steam} \\ \text{condensing}\end{matrix} + \begin{matrix}\text{heat lost by} \\ \text{condensed steam} \\ \text{cooling}\end{matrix} = \begin{matrix}\text{heat gained} \\ \text{by water}\end{matrix}$$

Let the mass of steam be m kg

$$m \times 2257 + m \times 4.187\,(100-70) = 100 \times 4.187\,(70-15)$$
$$2257m + 125.6m = 23\,028.5$$
$$2382.6m = 23\,028.5$$

$$m = \frac{23\,028.5}{2382.6} = 9.67 \text{ kg}$$

Example 8.11

Steam jets are used to remove 500 kg of ice at 0°C from the freezing chamber of an ice plant under repair. The effluent water is found to be at 10°C. How much steam at 100°C is used?

Solution

Latent heat lost by steam condensing + heat lost by condensed steam cooling

= latent heat required to melt ice + heat gained by melted ice

Let the mass of steam be m kg

$$m \times 2257 + m \times 4.187(100-10) = 500 \times 335 + 500 \times 4.187(10-0)$$
$$2257m + 376.8m = 167\,500 + 20\,935$$
$$2633.8m = 188\,435$$

$$m = \frac{188\,435}{2633.8} = 71.5 \text{ kg}$$

8.8 CHANGE OF SIZE

Many substances are affected dimensionally by changes in temperature. This is of vital importance in engineering where dimensional accuracy is so critical; also it is an effect which can be very useful to the engineer.

8.8.1 Thermal Expansion

Earlier we noted that the vibrating molecules of a solid move more vigorously as their temperature is increased. This produces more violent collisions with their neighbours so that the 'elbow room' of each molecule is increased. As a result of each molecule occupying more space in this way, the over-all length and breadth of the material are increased slightly—we say it has *expanded* (see figure 8.9). This effect is used in some engineering applications to alter the size of components by heating so they can be assembled easily but then fit very tightly together at normal temperature. For example, the gudgeon pin in a motorcycle piston is difficult to insert at room temperature but if the piston is first placed in hot water, to make the hole increase in diameter, the pin may be inserted more easily. In other types of work a bearing can be chilled in liquid nitrogen at −196°C to make it shrink until it will slide easily into its socket. The practice of heating the rim (or tyre) of a locomotive wheel to fit it into position is well known; when cool the tyre grips the wheel very tightly without the need for welding or other fixings.

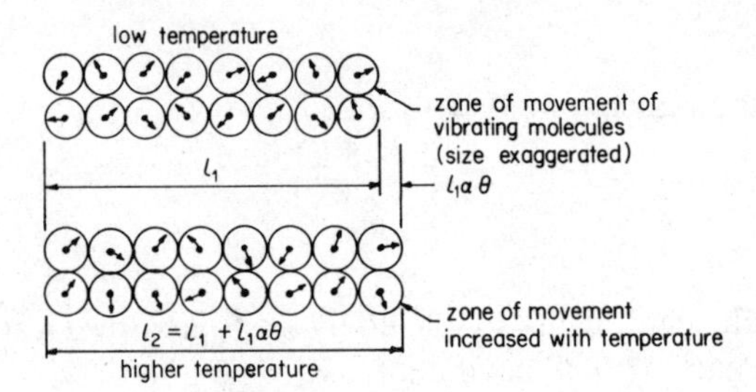

Figure 8.9

8.8.2 Measurement of Expansion

The change of length produced is *proportional* to the rise in temperature and to the *original length* of the object. We can understand this best if we think about the expansion of a chain—if one link expands by 1 mm, two links will expand by 2 mm, and so on; if we increase the length of the chain we shall also increase the amount of expansion produced. The third factor affecting the size of the expansion is the nature of the material itself. Different substances expand by different amounts and we define the *coefficient of linear expansion*, symbol α, as the increase in length per unit length per unit temperature rise where

$$\alpha = \frac{\text{change in length}}{\text{original length} \times \text{temp. change}}$$

therefore

$$\text{change in length} = l_0 \alpha \theta$$

and

$$\begin{aligned}\text{final length} &= \text{original length} + \text{change in length} \\ &= l_0 + l_0 \alpha \theta \\ &= l_0(1 + \alpha\theta)\end{aligned}$$

(see figure 8.9.)

Example 8.12

A pendulum used in a large clock is 2 m long. Calculate the change in length if the temperature rises by 12 K and the pendulum is made of steel ($\alpha_{steel} = 0.000\,015$/K).
Solution

$$\begin{aligned}\text{Change in length} &= l_0 \alpha \theta \\ &= 2 \times 0.000\,015 \times 12 \\ &= 0.00036 \text{ m} \\ &= 0.36 \text{ mm}\end{aligned}$$

Example 8.13

A shaft is 12.01 mm in diameter and has to be inserted in a hole exactly 12 mm in diameter. By how much must it be cooled down to make this possible without the use of force? ($\alpha = 0.000018$/K)
Solution

$$\text{Change in length} = l_0 \alpha \theta$$
$$0.01 = 12.01 \times 0.000018 \times \theta$$

$$\theta = \frac{0.01}{12.01 \times 0.000018} = 46.3 \text{ K}$$

Thus the shaft must be cooled down by 46.3 K.

8.8.3 Change of Surface Area

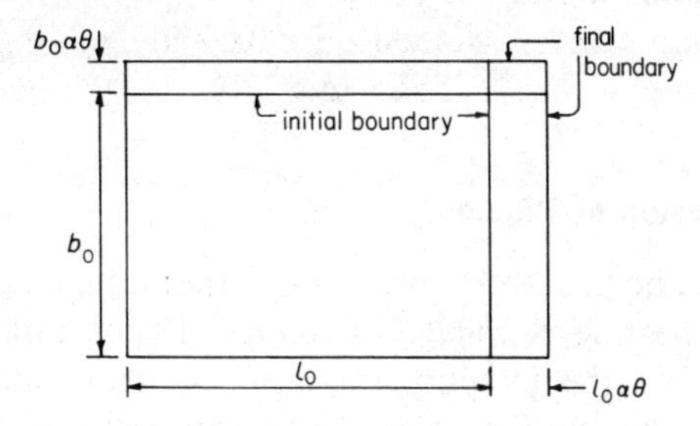

Figure 8.10

Consider a surface heated by θ K

$$\begin{aligned}\text{new length} &= l_0(1 + \alpha\theta) \\ \text{new breadth} &= b_0(1 + \alpha\theta) \\ \text{new area} &= l_0(1 + \alpha\theta) \times b_0(1 + \alpha\theta) \\ &= l_0 b_0(1 + 2\alpha\theta + \alpha^2\theta^2)\end{aligned}$$

Now α^2 is very small and can be ignored, thus

$$\text{new area} = \text{original area } (1 + 2\alpha\theta)$$

$$\text{change in area} = \text{original area} \times 2\alpha\theta$$

The value 2α is the *coefficient of superficial expansion* of the substance, symbol β, thus $\beta = 2\alpha$.

8.8.4 Change of Volume

A similar proof can be used to show that the *coefficient of cubical expansion* is 3α, symbol γ.

Example 8.14

Calculate the increase in volume of a rectangular block of copper

30 mm × 15 mm × 10 mm when it is heated from 10°C to 250°C. ($\alpha = 0.000017$/K)
Solution

$$\begin{aligned}\text{Initial volume } V_0 &= 30 \times 15 \times 10 = 4500 \text{ mm}^3\\ \text{Change in volume} &= V_0 \times 3\alpha \times \theta\\ &= 4500 \times 3 \times 0.000017 \times 240\\ &= 55.08 \text{ mm}^3\end{aligned}$$

8.8.5 Expansion of Liquids

Liquids have no 'length' or 'breadth', so the only change that takes place with heating is *change of volume*. This is utilised in such instruments as the mercury thermometer. For calculating the change we use a coefficient of *cubical* expansion, γ, which is not related to other values.

Example 8.15

Calculate the increase in volume of 2 m^3 of water when heated from 15°C to boiling point if the value of γ is 0.00021/K.
Solution

$$\begin{aligned}\text{Change in volume} &= 2 \times 0.00021\ (100 - 15)\\ &= 0.0357 \text{ m}^3 = 35.7 \text{ litres}\end{aligned}$$

8.9 PROPERTIES OF GASES

The special characteristics of gases require a more detailed treatment than solids and liquids, especially since they are used extensively in heat-conversion machines (engines).

8.9.1 Expansion of Gases

A gas is matter in its *freest state* and may be compressed and expanded very easily when compared with other states of matter. Gases always fill their container *completely*, irrespective of pressure or temperature because their molecules can move about without restriction (figure 8.11). To understand the behaviour of a gas when heated we must examine the relationship between temperature, volume *and* pressure. These are easy to follow if we consider how one property affects another when the *third is kept constant*.

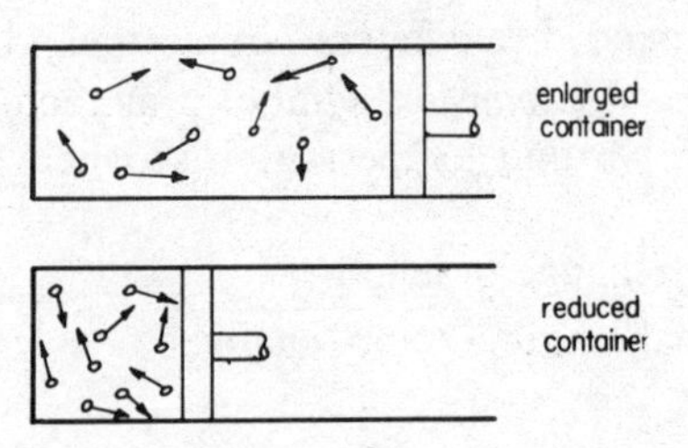

Figure 8.11 When gas is compressed, intermolecular space is reduced

Note on Pressure Measurement

Pressure is defined as force per unit surface area so it will be measured in units of N/m^2. 1 N/m^2 has been likened to 'the weight of an apple spread over a desk-top'—a very low value of pressure. For this reason we often state pressures in kN/m^2, for example, atmospheric pressure is 100 kN/m^2. The *bar* is an acceptable non-standard unit which is used for convenience: 1 bar = 100 kN/m^2 = 10^5 N/m^2. Changes in atmospheric pressure are sometimes quoted in millibars.

8.9.2 Boyle's Law

Formulated by Robert Boyle,[3] this relates pressure and volume when the temperature is unchanged. The pressure of a given mass of gas is *inversely proportional* to the volume provided the temperature remains constant. In symbol form

$$P \propto \frac{1}{V} \quad \text{or } P = k \times \frac{1}{V} \quad \text{or } PV = k$$

where k is a constant of proportionality. This means, for example, that *doubling* the pressure *halves* the volume (figure 8.12).

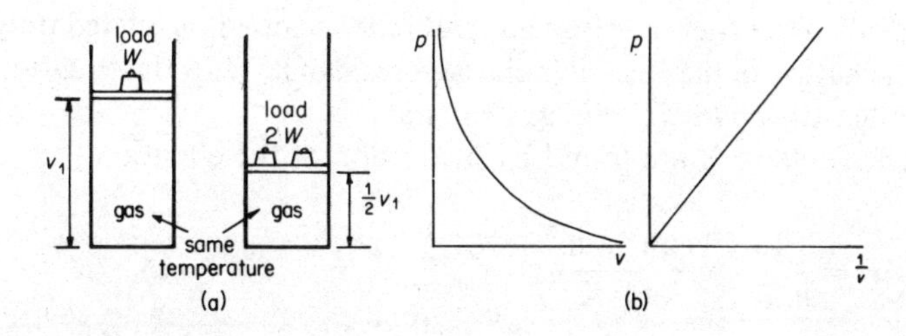

Figure 8.12

Example 8.16

A gas occupies 2 m^3 at a pressure of 200 kN/m^2. If the pressure is reduced to 150 kN/m^2 will the volume be increased or decreased if the temperature is kept constant, and by how much?
Solution $PV = k$; the initial values are P_1 and V_1 and final values P_2 and V_2, therefore

$$P_1V_1 = P_2V_2$$
$$200 \times 2 = 150 \times V_2$$

$$V_2 = \frac{200 \times 2}{150} = 2.67\ m^3$$

The volume has increased by 0.67 m^3.

8.9.3 Charles's Law

Formulated by Jacques Charles,[4] this relates volume and temperature when the pressure is kept constant. The volume of a given mass of gas is *directly proportional* to the *absolute* temperature.

$$V \propto T \quad \text{or } V = kT \quad \text{or } \frac{V}{T} = k$$

where k is a constant of proportionality (figure 8.13).

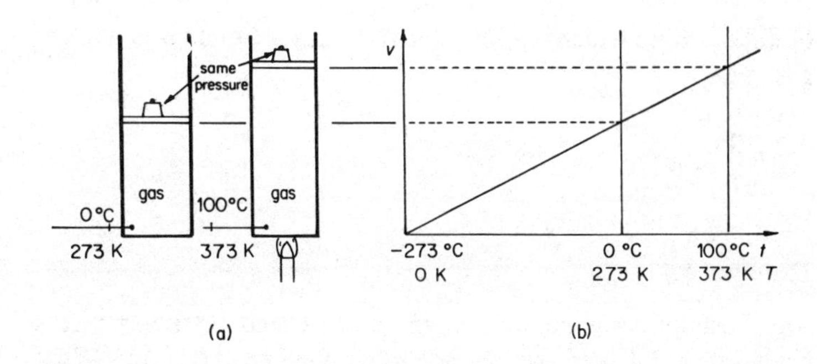

Figure 8.13

8.9.4 The Pressure Law

This relates pressure and temperature if *volume is kept constant*. The pressure of a given mass of gas is *directly proportional* to the absolute temperature.

$$P \propto T \quad \text{or } P = kT \quad \text{or } \frac{P}{T} = k$$

where k is a constant of proportionality (figure 8.14).

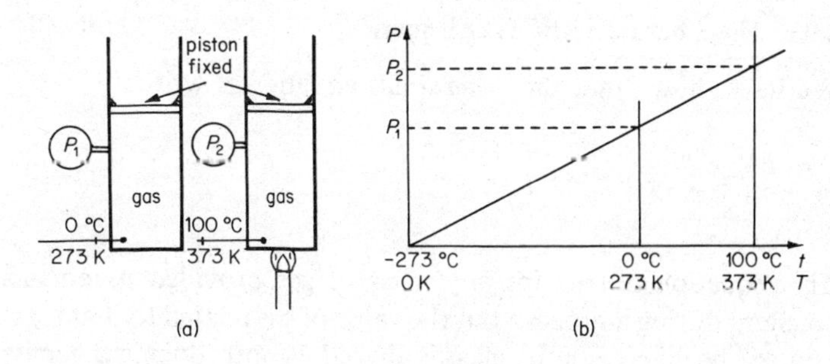

Figure 8.14

8.9.5 The General Gas Equation

We can combine the three relationships above into a single, very useful equation which is applicable to *all* gases. For a given mass of gas

$$\frac{PV}{T} = k$$

Example 8.17

A gas occupies a volume of 0.1 m^3 at 27°C and 100 kN/m^2. It is heated to 177°C and compressed into a volume of 0.05 m^3. What will be its new pressure?
Solution (Remember to use Kelvin temperatures.) Original conditions are P_1, V_1, T_1, final conditions $P_2 V_2$ and T_2. Then

$$\frac{P_1 V_1}{T_1} = k = \frac{P_2 V_2}{T_2}$$

$$\frac{100 \times 0.1}{(27+273)} = \frac{P_2 \times 0.05}{(177+273)}$$

$$P_2 = \frac{100 \times 0.1 \times 450}{0.05 \times 300} = 300 \text{ kN/m}^2$$

8.9.6 The Characteristic Gas Equation

We have shown that the *general* gas equation is written

$$\frac{PV}{T} = k \text{ or } PV = kT$$

This equation is true for any mass of gas provided it remains constant during a process, but the value of k is related to that mass and will be different if the mass is altered. By introducing the mass of gas m into the equation we find that a constant is obtained which is valid for *any* mass of a particular gas.

$$PV = kT$$
$$PV = mRT$$

R is called the *characteristic gas constant* because it is related only to the nature of the gas—it is characteristic of it. Thus the equation is called the *characteristic gas equation.*

The units of R are found by inspection of the equation

$$R = \frac{PV}{mT} = \frac{\text{N/m}^2 \times \text{m}^3}{\text{kg} \times \text{K}}$$

$$= \frac{\text{N m}}{\text{kg K}} = \frac{\text{J}}{\text{kg K}}$$

Typical values of R are: for air, 287 J/kg K; for carbon dioxide, 189 J/kg K; for hydrogen, 4.16 kJ/kg K.

Example 8.18

Calculate the volume occupied by 2 kg of air at 30°C and 1 bar pressure (10^5 N/m^2).
Solution

$$PV = mRT$$

$$V = \frac{mRT}{P}$$

$$= \frac{2 \times 287 \times (30+273)}{10^5}$$

$$= 1.74\ m^3$$

Example 8.19

Calculate the mass of carbon dioxide at 12 bars pressure in a cylinder of compressed gas at 15°C if the volume is 0.05 m^3.
Solution

$$PV = mRT$$

$$m = \frac{PV}{RT}$$

$$= \frac{12 \times 10^5 \times 0.05}{189 \times (15 + 273)}$$

$$= 1.1 \ kg$$

(Note that pressures used are in N/m^2 and temperatures are in K.)

1 Anders Celsius (1701–44) was a Swedish astronomer, professor at Uppsala University. He devised the 'centigrade' or 'hundred steps' temperature scale in 1742. The name of the scale was changed to Celsius in 1948.
2 Lord Kelvin (born William Thomson) (1824–1907) was a notable Scottish scientist, carrying out fundamental work on electricity and thermodynamics. He was the first to recognise the existence of an absolute zero of temperature (1848). He also achieved distinction with his interpretation of the work of Sadi Carnot.
3 Robert Boyle (1627–91) born in Ireland, was an enthusiastic pioneer of scientific experiment. He discovered (1662) the relationship between the pressure and volume of a gas at constant temperature, and conceived the theory that gases consist of discrete particles, separated by a void. Boyle also conducted extensive investigations in the field of chemistry. He was elected President of the newly formed Royal Society. Boyle cooperated with Robert Hooke (see p. 15).
4 Jacques A. C. Charles (1746–1823) was a French scientist and a teacher at the Sorbonne. He investigated expansion of gases with temperature, discovering that all gases expanded at the same rate if the pressure was constant (1787). This contributed to the concept of an absolute zero, postulated by Kelvin. Among his many other distinctions, Charles was the first to advocate the use of hydrogen for balloons.

TO THE STUDENT

At the end of this chapter you should be able to

(1) recognise that heat is a form of energy which can produce certain effects
(2) understand the methods by which heat is transferred from one place to another
(3) measure the effects of heat by temperature change
(4) calculate heat quantities involved in thermal processes
(5) understand the effects of heat on change of phase and dimensions
(6) understand the special laws governing the properties of gases
(7) complete exercises 8.1 to 8.6.

EXERCISES

8.1 In what way is heat *mainly* emitted from (a) a domestic 'radiator', (b) the Sun and (c) an electric fire element embedded in a silica casing.

8.2 Is it true to say that (a) a thermometer measures the heat in a body; or (b) a thermometer can be used to provide information from which the heat in a body can be found?

8.3 (a) What is the ratio of size between 1 degree centigrade, 1 degree Fahrenheit and 1 kelvin?
(b) Convert a temperature of 65°F (industrial working conditions) to its equivalent in °C. [*Hint*. Since −40°F = −40°C we can follow this sequence: (i) add 40 to temperature, (ii) convert to other size of degree, (iii) subtract 40 from the result. This method works in either direction.] (c) What is 65°F expressed on the Kelvin scale?

8.4 10 kg of water at 0°C are contained in a copper vessel of mass 6 kg. Another 10 kg of boiling water at 100°C are added to the first amount. (a) Why is the final temperature unlikely to be mid-way between 0°C and 100°C? (b) Calculate its value. (c) Would the final temperature be the same if the vessel held ice at 0°C to begin with? Justify your answer.

8.5 (a) Calculate the value of the coefficient of expansion of a metal bar if the length increases from 800 mm to 801.44 mm when the temperature changes from 0°C to 100°C. (b) What metal do you think the bar is made of? (c) What is the percentage increase in its surface area?

8.6 0.005 kg of air is compressed in an engine cylinder. The initial pressure is 10^5 N/m^2 and the initial temperature 300 K (27°C). (a) Calculate the initial volume of the cylinder. The volume is now reduced to 1/10 of its original size by the movement of the piston. Calculate the final pressure if (b) the compression is so slow that the temperature stays at 300 K; (c) the compression is rapid and the final temperature is 600 K (sometimes referred to as 'heat of compression').

NUMERICAL SOLUTIONS

8.3 (a) F:C:K = 1:1.8:1.8, (b) 18.3 °C, (c) 291.3 K
8.4 (b) 48.65 °C, (c) no
8.5 (a) 0.000 018/K, (c) 0.0036%
8.6 (a) 0.0043 m^3, (b) 10^6 N/m^2, (c) 2×10^6 N/m^2

9 Electricity and Electrical Measurement

The object of this chapter is to introduce the student to the basic concepts of current electricity and associated circuitry, and to relate these to engineering applications.

9.1 THE NATURE OF ELECTRIC CHARGE

Electric charge, like many physical phenomena, is invisible and therefore difficult to impress on the imagination, but we can recognise its *effects* quite readily and so build up a picture which explains the nature of electricity in a reasonably satisfactory way.

One effect easily demonstrated is the *force* exerted by an electric charge. Certain substances, such as ebonite, when rubbed with a piece of fur acquire the ability to repel each other with a small but detectable force. Thus the ebonite now has a property which it did not possess originally and we call this property electric charge. The natural material amber shows a similar effect and its Greek name, *elektron*, gives us our name for electricity.

Other materials, such as glass rubbed with silk, also acquire a charge, but this charge appears to behave in a somewhat different way from the ebonite/fur variety. For instance it *attracts* a charged piece of ebonite instead of repelling it. Experiments show that there are two types of electric charge: *negative* (ebonite/fur) and *positive* (glass/silk).

By experiment we can show that two positive charges close together repel each other, similarly two negative charges also repel; but when opposite charges (that is, one negative and one positive) are used the mutual force is attractive (see figure 9.1). Modern thinking relates the behaviour of charged bodies to the structure of

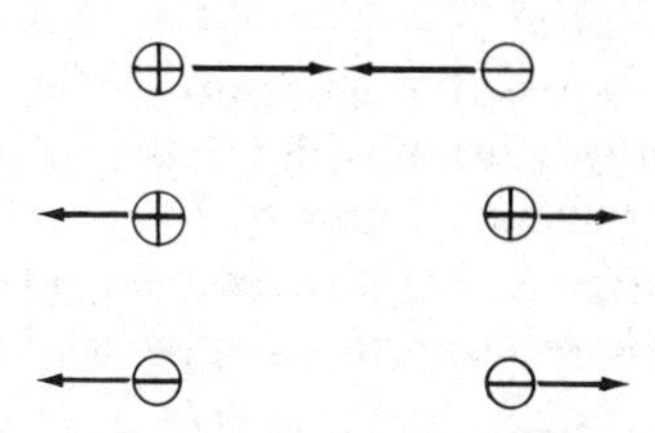

Figure 9.1 Forces between charges—unlike charges attract; like charges repel

the atom which is believed to consist of two main parts: a nucleus containing positive *protons* (plus neutral neutrons in some cases) and negative *electrons* in orbit around the nucleus (see figure 9.2). The atom is electrically neutral because the electrons and protons are present in equal numbers. Friction produced by rubbing may detach some electrons and produce a surplus of positive charge. On the other hand, rubbing with different materials may deposit some electrons (from the rubbing material) thus giving a surplus of negative charge. Experimental work has shown that the smallest charge ever detected is the charge on an electron, which confirms the argument that electron transfer is responsible for the existence of electric charge.

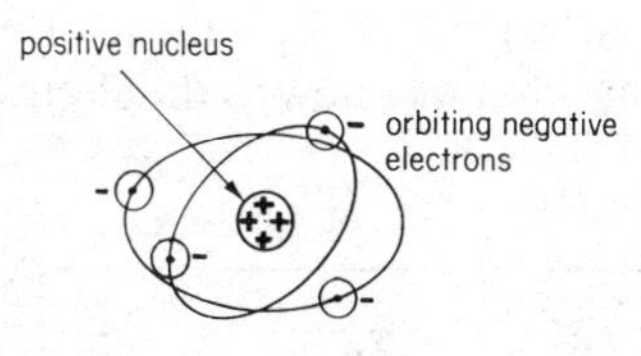

Figure 9.2 Simplified model of the atom

9.1.1 The unit of charge

The ideal unit of charge would be that of a single electron, but this is inconveniently small and a larger unit is used, called the *coulomb*.[1] This is the charge on a very large number of electrons (6.3×10^{18}), but it is actually defined in quite a different way, by considering the *rate of flow* of the charge along a conductor.

Charges are able to pass through many materials and in doing so they dissipate energy. The rate of flow is easy to detect and measure with suitable instruments; a flow of *1 coulomb per second* being referred to as *1 ampere* of electric *current.*[2] (The ampere itself is defined from its *electromagnetic effect*, which is fully described in chapter 10.)

9.2 THE ELECTRIC CURRENT

All matter is composed of atoms which have orbiting electrons in their structure; but conducting materials, usually metals, include one or more electrons on their outermost orbits which are less firmly attached to the atom than the remaining electrons. These so-called *'free' electrons* migrate from atom to atom in the conductor, moving in all directions, but not leaving the metal. We can imagine the metal contains an 'electron cloud' which is uniformly distributed and in constant, but random, motion.

Now we have noted in section 9.1 that like charges attract and unlike charges repel, so application of positive and negative charges to opposite ends of the conductor will tend to move the free electrons (which are negatively charged) towards the positively charged end and away from the negative end. This is called *electron drift* and results in a general movement of electrons along the conductor, forming what we know as the electric current (see figure 9.3).

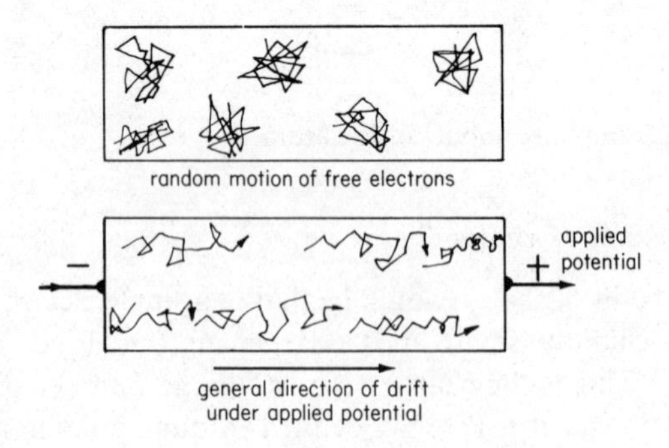

Figure 9.3

Materials which possess few free electrons are unable to pass any appreciable current and are called *insulators.* Most non-metallic solids are insulators, with carbon as a notable exception. There is an intermediate class of materials with special properties of charge transfer, called *semiconductors*, but these will not be considered here.

It will be noticed that the flow of current is from *negative to positive* and the basis is *electron flow*. In earlier times this was not clearly understood and the direction was arbitrarily declared to take place the opposite way, from positive to negative. For convenience this is still assumed to be the case and is called *conventional flow* (see figure 9.4).

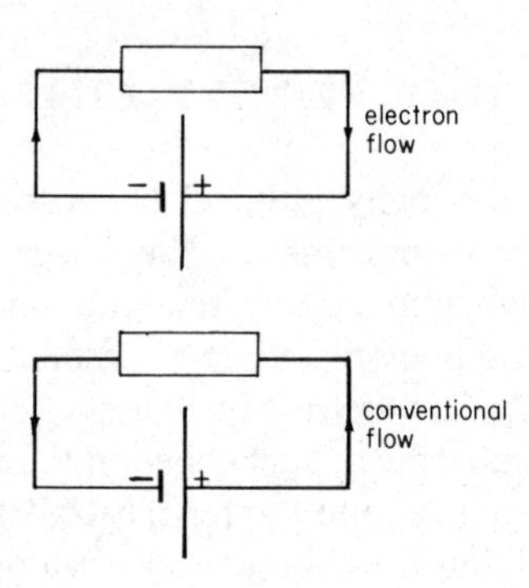

Figure 9.4

9.2.1 Electric Potential and Electromotive Force (e.m.f.)

The forces exerted between electric charges mean that as two similar charges are moved towards each other a quantity of work is done against the force and this energy is stored by the charged system, which is now 'pressurised' and tends to release the energy if allowed to do so. This 'pressure' is given the name *electric potential* and is measured in *volts.* If the energy stored by the charged system is 1 joule per coulomb of charge the potential is *1 volt.*[3]

Example 9.1

20 J of energy are used to transmit 5 C of electricity along a conductor. Calculate the potential difference between the ends of the conductor.
Solution

$$\text{p.d. (volts)} = \frac{\text{J}}{\text{C}}$$

$$= \frac{20}{5} = 4 \text{ V}$$

The concept of a flow of charge or electric current has been introduced, but it should be noted that current will not flow unaided and always requires a *potential difference* (p.d.) or pressure difference to promote the flow. This is called an *electromotive force* or *e.m.f.* since it is responsible for the motion of the electric charge. (The word force is used loosely and should *not* be interpreted in the same way as a mechanical force; it is better to use the expression *e.m.f.*) The units of e.m.f. are *volts*. Various methods are used to produce an e.m.f., including *cells* and batteries (see section 9.9) and *generators* (see section 10.6). These are the most widely used although other methods, such as the photoelectric effect and thermocouple, are used in special applications.

9.3 THE ELECTRICAL CIRCUIT

To maintain a continuous flow of current through a conductor it must be connected in a *closed loop* which includes the source of e.m.f. If the circuit were not a closed loop, the charge would accumulate at some point and develop its own potential in opposition to the e.m.f., preventing further flow. Such a closed loop is a *circuit* (see figure 9.5). The conductors forming a circuit impede the flow of current but do not prevent it; this property is called *resistance* and is caused by the need to move the free electrons away from their original location against the force of attraction which originally kept them there. As a result of this disruption, energy is

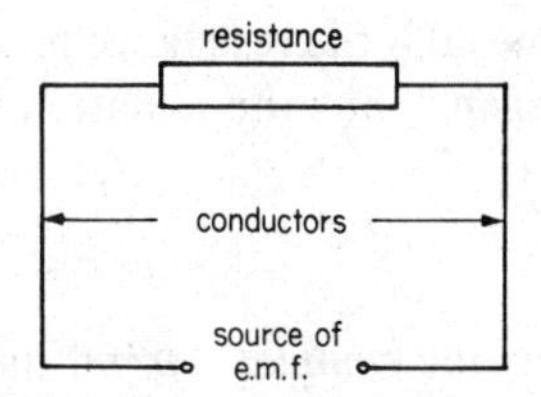

Figure 9.5 Simplified electrical circuit

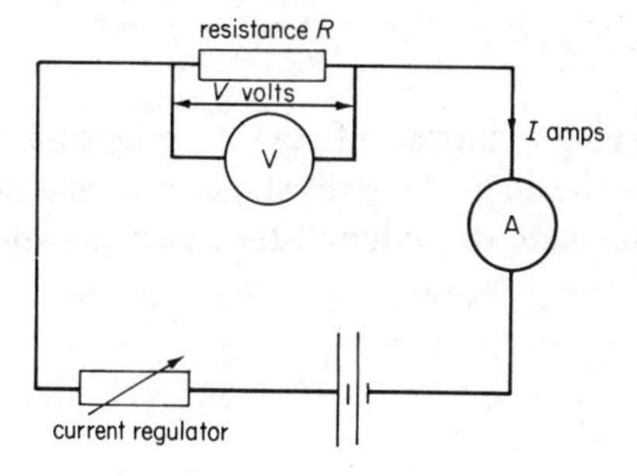

Figure 9.6 Demonstration of Ohm's law

released, often in the form of heat, and a *reduction in potential* takes place. The relationship between this potential 'drop' and the current flowing is embodied in *Ohm's*[4] *law*, which states that the potential difference across the ends of the conductor is directly proportional to the current flowing in the conductor. This can be verified with a simple experiment, as shown in figure 9.6. If the voltmeter reading is V volts (representing the potential difference) and the ammeter reading is I amperes (the current), then

$$V \propto I$$

or

$$V = I \times \text{a constant}$$

This constant is the resistance R measured in ohms, where *1 ohm* (symbol Ω, Greek omega) is the resistance producing a p.d. of *1 volt* with a current of *1 amp*. Thus the equation becomes

$$V = IR$$

If the value of R is for the *complete circuit* the potential difference across it will be equal to the e.m.f. E so the equation becomes

$$E = IR$$

Example 9.2

A circuit carrying a current of 2.35 A contains a resistance. The voltage across the ends of the resistance is measured and found to be 6.55 V. Calculate the value of the resistance in Ω.
Solution

$$V = IR$$

$$R = \frac{V}{I}$$

$$= \frac{6.55}{2.35} = 2.79\Omega$$

Example 9.3

Calculate the current flowing in a resistance of 15 Ω when connected to a source of 24 V.
Solution

$$V = IR$$

$$I = \frac{V}{R}$$

$$= \frac{25}{15} = 1.67 \text{ A}$$

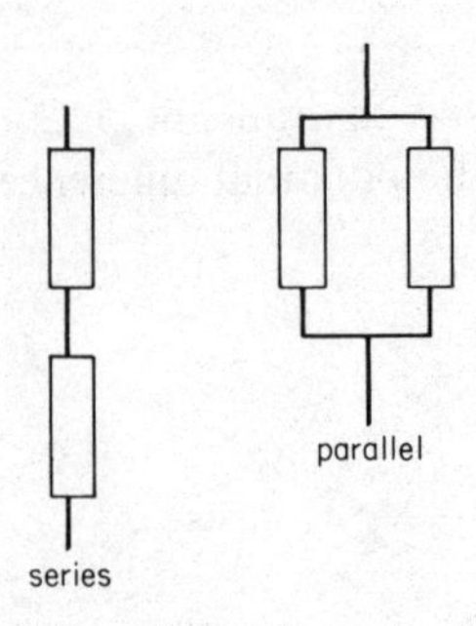

Figure 9.7 Combined resistances

9.3.1 Combined Resistances

A circuit may be composed of several conductors called *resistors* joined together *in series* (end to end) or *in parallel* (side by side)—see figure 9.7. The total resistance of the circuit is called the *combined resistance* and depends on the system of connections used.

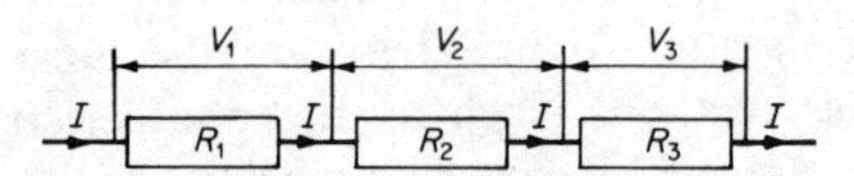

Figure 9.8 Series connection

Series Connection (*figure* 9.8)

The same current, I, flows through each resistance so the respective p.d.s are

$$V_1 = IR_1 \quad V_2 = IR_2 \quad V_3 = IR_3$$

Therefore

$$\text{total p.d.} = V_1 + V_2 + V_3 = V_{\text{total}}$$

and

$$V_{total} = IR_1 + IR_2 + IR_3$$
$$= I(R_1 + R_2 + R_3)$$

But $V_{total} = IR_{combined}$, therefore for series connection

$$R_{combined} = R_1 + R_2 + R_3$$

Example 9.4

Resistances of 6 Ω, 12 Ω and 25 Ω are connected in series. (a) What is their combined resistance? (b) What is the percentage change in value when a further resistance of 2.5 Ω is connected in series?

Solution

(a)

$$R_{combined} = R_1 + R_2 + R_3$$
$$= 6 + 12 + 25 = 43\ \Omega$$

(b)

$$\text{New resistance} = 43 + 2.5 = 45.5\ \Omega$$

$$\%\ \text{change} = \frac{2.5}{43} \times 100 = 5.8\%$$

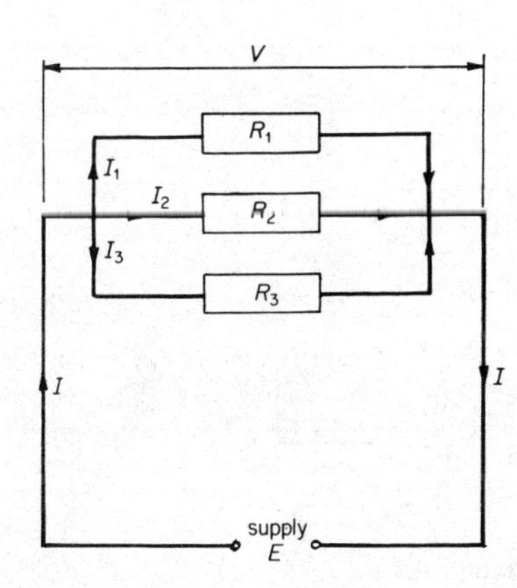

Figure 9.9 Parallel connection

Parallel Connection (figure 9.9)

The total current is divided between the different resistors, but each resistor has the same p.d., V volts, across its ends, thus

$$I_1 = \frac{V}{R_1} \qquad I_2 = \frac{V}{R_2} \qquad I_3 = \frac{V}{R_3}$$

therefore

$$\text{total current } I = I_1 + I_2 + I_3$$
$$= \frac{V}{R_1} + \frac{V}{R_2} + \frac{V}{R_3}$$
$$= V\left(\frac{1}{R_1} + \frac{1}{R_2} + \frac{1}{R_3}\right)$$

hence

$$\frac{I}{V} = \left[\frac{1}{R_1} + \frac{1}{R_2} + \frac{1}{R_3}\right]$$

Now for this circuit $V = E$, the e.m.f., and since $E = IR_{combined}$

$$\frac{I}{V} = \frac{1}{R_{combined}}$$

therefore for parallel connection

$$\frac{1}{R_{combined}} = \frac{1}{R_1} + \frac{1}{R_2} + \frac{1}{R_3}$$

Example 9.5

Resistances of 12 Ω, 14 Ω and 16 Ω are connected in parallel. (a) Find the combined resistance in the circuit. (b) Calculate the percentage change in resistance when a further resistance of 10 Ω is added in parallel with the others.

Solution

(a) $$\frac{1}{R}=\frac{1}{R_1}+\frac{1}{R_2}+\frac{1}{R_3}$$

$$=\frac{1}{12}+\frac{1}{14}+\frac{1}{16}$$

$$=0.0833+0.0714+0.0625=0.2172$$

therefore

$$R=\frac{1}{0.2172}=4.6\ \Omega$$

(b) $$\frac{1}{R_{\text{total}}}=\frac{1}{4.6}+\frac{1}{10}$$

$$=0.2174+0.1=0.3174$$

therefore

$$R_{\text{total}}=\frac{1}{0.3174}=3.15\ \Omega$$

$$\%\ \text{change}=\frac{4.6-3.15}{4.6}\times 100=31.52\%\ \text{(reduction)}$$

Note that for parallel connection, the combined resistance is always smaller than the smallest component resistance of the system.

A quick method for finding the combined resistance of *two* resistors in parallel is to use the formula

$$R_{\text{combined}}=\frac{\text{product}}{\text{sum}}$$

Since

$$\frac{1}{R_{\text{combined}}}=\frac{1}{R_1}+\frac{1}{R_2}=\frac{R_1+R_2}{R_1R_2}$$

hence

$$R_{\text{combined}}=\frac{R_1R_2}{R_1+R_2}$$

A complex system of resistors is often referred to as a *network*.

Example 9.6

Calculate the combined resistance of the following pairs of resistors in parallel. (a) 2 Ω, 2 Ω, (b) 6 Ω, 3 Ω, (c) 2.5 Ω, 10 Ω, (d) 12 Ω, 5 Ω.

Solution

(a)

$$R_c=\frac{2\times 2}{2+2}=\frac{4}{4}=1\ \Omega$$

(b)

$$R_c=\frac{6\times 3}{6+3}=\frac{18}{9}=2\ \Omega$$

(c)

$$R_c=\frac{2.5\times 10}{2.5+10}=\frac{25}{12.5}=2\ \Omega$$

(d)

$$R_c=\frac{12\times 5}{12+5}=\frac{60}{17}=3.53\ \Omega$$

9.3.2 Conductance

The reciprocal of resistance is referred to as the *conductance* G of a conductor, where

$$G = \frac{1}{R}$$

The unit is the *siemens*[5] (plural also siemens, symbol S) and the quantity represents the readiness to permit a current to flow, that is, current is directly proportional to G. A low conductance permits only a low current and vice versa.

Example 9.7

(a) Calculate the conductance of the following resistances (i) 5Ω (ii) 16Ω (iii) 25Ω. (b) Hence find their combined conductance when connected in parallel.

Solution
(a)

$$G_1 = \frac{1}{5} = 0.2 \text{ S}$$

$$G_2 = \frac{1}{16} = 0.0625 \text{ S}$$

$$G_3 = \frac{1}{25} = 0.04 \text{ S}$$

(b) for parallel connection

$$\begin{aligned} G_{\text{combined}} &= G_1 + G_2 + G_3 \\ &= 0.2 + 0.0625 + 0.04 \\ &= 0.3025 \text{ S} \end{aligned}$$

9.3.3 Series – Parallel Combinations of Resistances

To find the combined resistance of such networks the resistance of its component parts is determined, stage by stage, until the whole network is reduced to a single resistance (figure 9.10). More

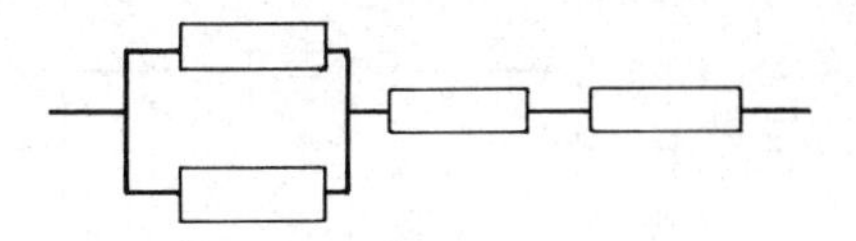

Figure 9.10 Series–parallel combination

complex networks are analysed by other methods which require study beyond the scope of this book.

Example 9.8

Calculate the combined resistance of the network shown in figure 9.11.

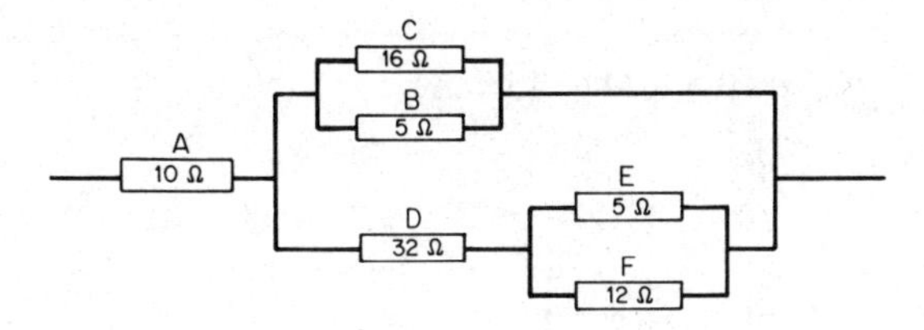

Figure 9.11

Solution Evaluate the combined resistances of the parallel pairs B and C and E and F.

$$R_{BC} = \frac{R_B R_C}{R_B + R_C} = \frac{16 \times 5}{16 + 5} = \frac{80}{21} = 3.81 \ \Omega$$

$$R_{EF} = \frac{R_E R_F}{R_E + R_F} = \frac{5 \times 12}{5 + 12} = \frac{60}{17} = 3.53 \ \Omega$$

The network is now simpler (figure 9.12). Evaluate the series connected pair D and EF

$$R_{DEF} = 32 + 3.53 = 35.53 \ \Omega$$

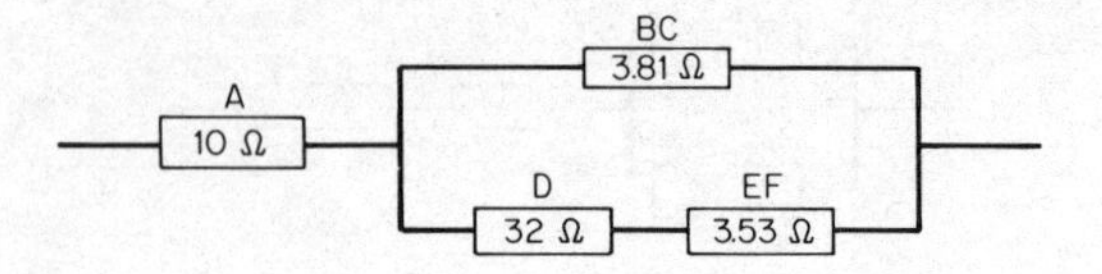

Figure 9.12

Next evaluate the parallel pair formed by BC and DEF (figure 9.13)

$$R_{BCDEF} = \frac{3.81 \times 35.53}{3.81 + 35.53} = 3.44\ \Omega$$

Finally evaluate the series pair A and BCDEF

$$R_{combined} = 10 + 3.44 = 13.44\ \Omega$$

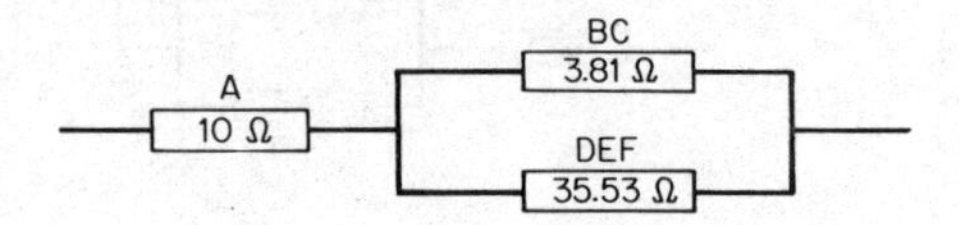

Figure 9.13

9.3.4 Calculation of Potential at Different Points in a Network

This requires a methodical application of Ohm's law and the following

total current entering network = total current leaving

total applied potential between two points = sum of p.d.s between the same two points

Example 9.9

If a potential of 15 V is applied across the ends of the network in example 9.8 calculate the potential at the points Y and Z in figure 9.14.

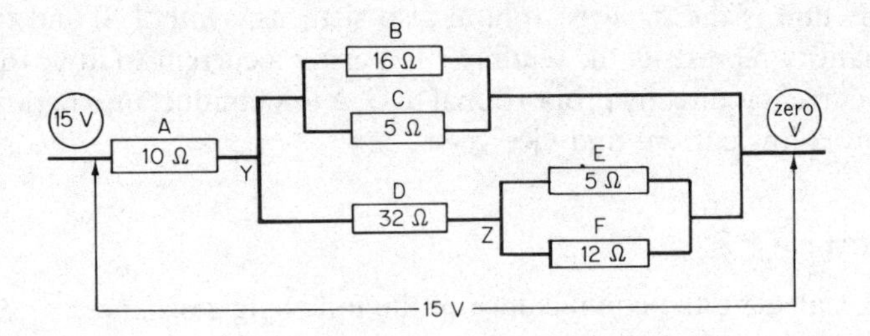

Figure 9.14

Solution Total current through network is given by

$$\frac{V}{R} = \frac{15}{13.44} = 1.12\ \text{A}$$

This is also the current through resistor A. Hence

$$\text{p.d. across A} = IR$$

$$= 1.12 \times 10 = 11.2\ \text{V}$$

therefore

$$\text{potential at Y} = 15 - 11.2 = 3.8\ \text{V}$$

Consider now the branch DEF which we know has a combined resistance of 35.53 Ω. Then

$$\text{current in this branch} = \frac{V}{R} = \frac{3.8}{35.53} = 0.107\ \text{A}$$

This is also the current flowing through D. Hence

$$\text{p.d. across D} = IR$$
$$= 0.107 \times 32 = 3.42 \text{ V}$$
$$\text{potential at Z} = 3.8 - 3.42 = 0.38 \text{ V}$$

Check: Total p.d. = sum of individual p.d.s (figure 9.15)

$$11.2 + 3.42 + 0.38 = 15 \text{ V}$$

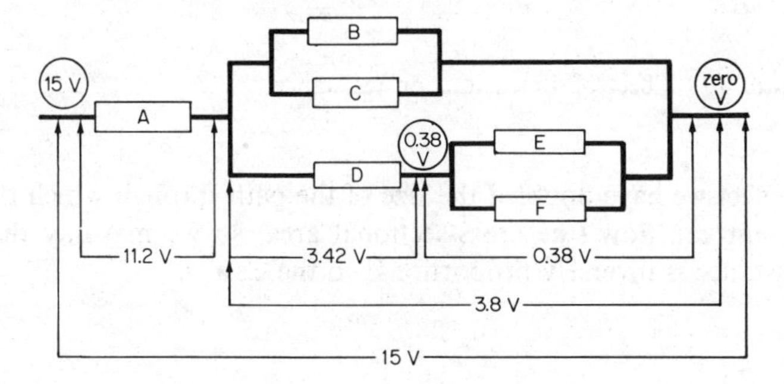

Figure 9.15

9.3.5 Internal Resistance of a Source of E.M.F.

If a cell or other source of e.m.f. is tested with a voltmeter, with no other components in circuit, the reading obtained is called the *open-circuit* voltage because no current is taken from the source (we can neglect the very small current through the voltmeter). If the cell is now connected to a circuit so that a current is delivered, the voltage reading falls to a lower value, the difference often being referred to as 'lost' volts. It is as if the circuit included an *extra resistance* which becomes connected inside the cell only when delivering a current to an external load. This resistance is called the *internal resistance* of the source (figure 9.16) and applies to all types of supply device including cells and generators. It has no effect when the cell is on open circuit since no potential drop will occur if the current is zero, thus

$$V = IR$$
$$= \text{zero} \times R = \text{zero V}$$

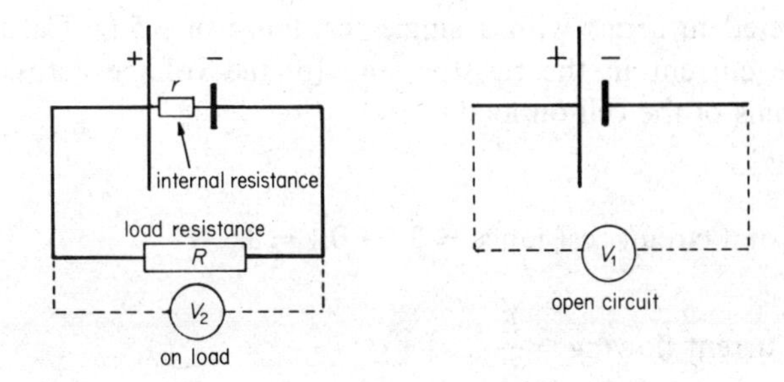

Figure 9.16

Let the open-circuit voltage be V_1 and the voltage on load be V_2. If the resistance in the external circuit is R Ω then the current taken from the source is

$$I = \frac{V_2}{R} \text{ A}$$

Now the lost volts $(V_1 - V_2)$ is the p.d. produced by I flowing through the internal resistance r. Hence

$$V_1 - V_2 = Ir$$

$$r = \frac{V_1 - V_2}{I}$$

It will be seen that the lost volts are not a constant quantity but depend on the current taken from the source. Substituting for I we have

$$r = \frac{(V_1 - V_2)R}{V_2}$$

This is a convenient method for determining the value of r.

Example 9.10

A cell has an e.m.f. of 1.5 V and an internal resistance of 0.1 Ω. It is

connected in series with a single resistance of 3.5 Ω. Calculate (a) the current in the resistor, and (b) the voltage across the terminals of the cell on load.
Solution

$$\text{Total circuit resistance} = 3.5 + 0.1 = 3.6\ \Omega$$

$$\text{Current flowing} = \frac{E}{R}$$

$$= \frac{1.5}{3.6} = 0.417\ \text{A}$$

$$\text{p.d. across load resistance} = IR = 0.417 \times 3.5 = 1.46\ \text{V}$$

This is the terminal voltage on load. (Lost volts = 1.5 – 1.46 = 0.04 V.)

9.4 FACTORS DETERMINING RESISTANCE

The resistance R of a conductor represents its opposition to the flow of current, but the actual value of R depends on certain physical characteristics of the conductor. Suppose we have a uniform conductor of resistance R_1 and length l_1 and we connect this in series with another similar conductor. The combined resistance $R = 2R_1$ and the total length is $2l_1$ (figure 9.17). Thus we can say that resistance is directly proportional to the length of the conductor.

$$R \propto l$$

If these two conductors are now connected in parallel the resistance will be

$$R = \frac{R_1 R_1}{R_1 + R_1} = \frac{R_1^2}{2R_1} = \frac{1}{2}R_1$$

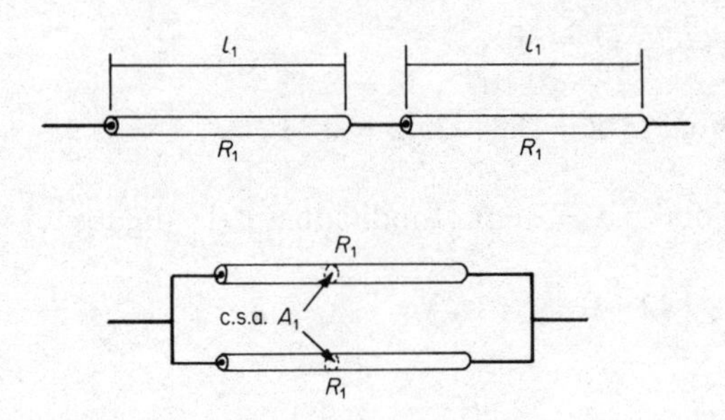

Figure 9.17 Effect of dimensions on resistance

In effect we have doubled the size of the path through which the current can flow (the cross-sectional area) so we may say that resistance is inversely proportional to the c.s.a. A

$$R \propto \frac{1}{A}$$

Combining the two relationships

$$R \propto \frac{l}{A}$$

or

$$R = \frac{l}{A} \times \text{a constant}$$

This constant depends on the type of material in use and is called the *resistivity* or *specific resistance* of the conducting material, symbol ρ. Thus

$$R = \rho\frac{l}{A}$$

or

$$\rho = \frac{RA}{l}$$

The units of ρ are ohm-metres. If a conductor were made of length 1 m and c.s.a. 1 m^2 the resistance would have the value ρ Ω.

Example 9.11

Calculate the diameter of a wire used to form a heating element 1.5 m long if the resistance is to be 60 Ω and the resistivity of the metal is 45×10^{-8} Ω m
Solution

$$R = \frac{\rho l}{A}$$

therefore

$$A = \frac{\rho l}{R}$$

$$= \frac{45 \times 10^{-8} \times 1.5}{60} = 1.125 \times 10^{-8} \text{ m}^2$$

$$= 0.01125 \text{ mm}^2$$

$$A = \frac{\pi d^2}{4}$$

therefore

$$d = \sqrt{\left(\frac{4A}{\pi}\right)} = \sqrt{\left(\frac{4 \times 0.01125}{\pi}\right)} = 0.12 \text{ mm}$$

9.4.1 Conductivity

This is the property of a material which indicates its readiness to conduct. Conductivity is thus the reciprocal of resistivity, symbol σ(sigma), and the units are siemens per metre.

Example 9.12

A 'third rail' power system for an electric train has a supply rail cross-sectional area of 800 cm^2. If the conductivity of steel is 9 $\times 10^6$ S/m calculate the resistance per km of the rail.
Solution

$$\text{R} = \frac{l}{\sigma A} = \frac{1000}{9 \times 10^6 \times 80 \times 10^{-4}}$$

$$= 1.39 \times 10^{-2} \, \Omega$$

9.4.2 Effect of Temperature on the Resistance of a Conductor

In the majority of materials (but not all) the effect of a temperature rise is to *impede* the flow of electrons through the material and the resistance is therefore increased. It can be shown that the change in the resistance is directly proportional to the temperature and also to the resistance at 0 °C (a convenient datum) R_0. Therefore

$$\text{change in resistance} \propto (\text{temperature } t\,^\circ\text{C}) \times R_0$$

or

$$\text{change in } R = R_0 t \times \text{a constant}$$

The constant is called the *temperature coefficient of resistance*, symbol α, where $\alpha = R/R_0 t$. The units are 1/kelvin, K^{-1} (see figure 9.18). Hence total resistance at temperature t is given by

$$R_t = R_0 + \text{change in resistance}$$
$$= R_0 + R_0 \alpha t$$

$$R_t = R_0(1 + \alpha t)$$

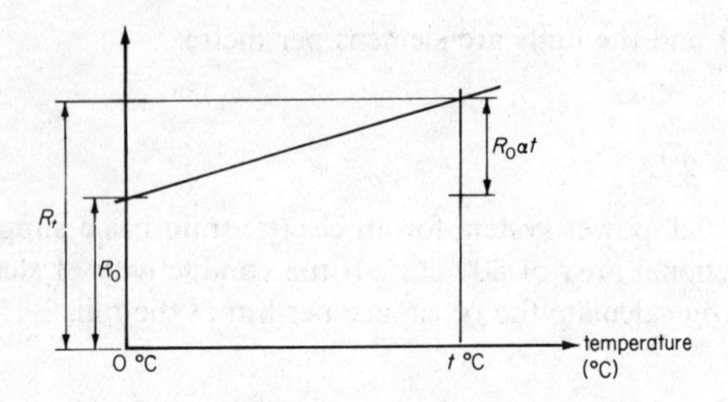

Figure 9.18 Temperature coefficient of resistance graph

Example 9.13

Calculate the resistance of a cable at 60°C if its resistance at 0°C is 2.45 Ω (α for the cable material = 5×10^{-3}/K).
Solution

$$R_{60} = R_0[1 + (\alpha \times 60)]$$
$$= 2.45[1 + (0.005 \times 60)]$$
$$= 3.185\ \Omega$$

The resistance at 0°C may not be known, but the resistance at temperature t can be found if the value at some different temperature *is* known. Suppose at the temperature t' the resistance is known to be R', we can then say

$$R' = R_0(1 + \alpha t')$$

Then at any other temperature t the resistance R_t is

$$R_t = R_0(1 + \alpha t)$$

Dividing

$$\frac{R_t}{R'} = \frac{R_0(1 + \alpha t)}{R_0(1 + \alpha t')}$$

Hence

$$R_t = R' \times \frac{(1 + \alpha t)}{(1 + \alpha t')}$$

Example 9.14

Calculate the resistance of a heater element at 200°C if its value at 100°C is 60 Ω and α is 50×10^{-4}/K.
Solution

$$R_{200} = R_{100} \times \frac{[1 + (\alpha \times 200)]}{[1 + (\alpha \times 100)]}$$

$$= 60 \times \frac{2.0}{1.5}$$

$$= 80\ \Omega$$

Special Materials

For the manufacture of resistors, materials are required which have a very low temperature coefficient of resistance, and at the same time possess a reasonably high resistivity. Various metals known as *resistance alloys* have been devised such as

constantan, $\alpha = \pm 0.4 \times 10^{-4}$/K
manganin, $\alpha = \pm 0.1 \times 10^{-4}$/K
nichrome, $\alpha = 1 \times 10^{-4}$/K

These temperature coefficient of resistance values should be compared with values for other metallic conductor materials (for example, α for copper = 43×10^{-4}/K).

The value of α for *carbon*, a material used as a conductor in special applications, is distinctive in being *negative*, that is, the resistance *decreases* as the temperature rises. This behaviour is also typical of some insulating materials, leading to a breakdown of insulation at very high temperatures.

9.5 ELECTRIC POWER

Power is defined as the *rate* of energy transfer or rate of doing work (section 7.9), and the flow of electricity through a resistance is responsible for the transfer of energy which is released as heat, or some other form (figure 9.19). The rate of energy transfer depends on the current and the potential drop across the resistance. Thus

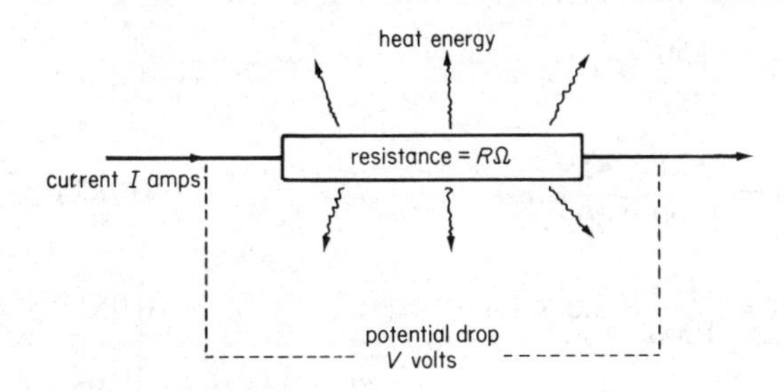

Figure 9.19 Power dissipation

$$\text{power} = \text{current} \times \text{voltage drop}$$

$$\text{current (amperes)} = \text{coulombs/second}$$
$$\text{p.d. (volts)} = \text{joules/coulomb}$$

hence

$$\text{power} = \frac{\text{coulombs}}{\text{second}} \times \frac{\text{joules}}{\text{coulomb}}$$

$$= \text{joules/second or } \textit{watts}^6$$

or

$$\text{power} = I \times V \text{ watts}$$

Alternatively, from Ohm's law we can say

$$V = IR$$

therefore

$$\text{power} = I \times IR$$
$$= I^2R \text{ watts}$$

Similarly

$$\text{power} = \frac{V^2}{R} \text{ watts}$$

Example 9.15

Calculate the power dissipated by a heater element of 60 Ω resistance if it carries a current of 4 A.
Solution

$$\text{Power} = I^2R$$
$$= 4 \times 4 \times 60$$
$$= 960 \text{ W}$$

9.5.1 Power Rating of Components

Resistors used in electrical equipment have to dispose of the heat generated within them. Usually the heat is transferred to the surrounding air, but this is limited by the maximum safe temperature of operation. This safe rate of heat flow to the atmosphere is equal to the power *rating* of the resistor. Resistors of insufficient power rating are liable to become damaged by overheating.

Example 9.16

Calculate the power rating required for a resistor of 47 kΩ which is to be connected across a supply of 30 V.
Solution

$$\text{Power} = \frac{V^2}{R} = \frac{30 \times 30}{47 \times 1000} = 0.019 \text{ W}$$

Typical resistors are available at 0.1 W rating, which is more than five times this value and quite satisfactory for the duty required.

9.5.2 Running Cost of Electrical Devices

The electricity supply industry charges its consumers on a basis of the *energy* used, the prices being determined by the cost of production and other commercial considerations. The unit of energy is of course the joule, but this is inconveniently small for most purposes. One alternative is the megajoule (MJ) and proposals have been made to adopt this for commercial use, but for the present the *kilowatt-hour* (kW h) is preferred. The kW h is defined as the energy consumed per hour by an appliance with a power of 1 kW. It can be seen that the kW h is equal to 3600 J or 3.6 MJ

$$\begin{aligned} 1\ \text{kW h} &= 1000\ \text{J/s} \times 3600\ \text{s} \\ &= 3.6 \times 10^6\ \text{J} \end{aligned}$$

For simplicity the kW h is often referred to as a '*unit*' of electricity when computing charges.

It should be noted that different rates are charged for a unit depending on various factors such as the quantity used, the *rate* of consumption, the time of day and the maximum demand made during the day. Details, which change from time to time, are available from the supply industry's offices.

Example 9.17

Calculate, to the nearest penny, the daily cost of an electric heating system rated at 4 kW if it operates for 11 hours each day and the cost per unit of electricity is 1.4 p.

Solution

$$\begin{aligned} \text{Number of units used} &= \text{power in kW} \times \text{time in h} \\ &= 4 \times 11 = 44\ \text{units} \\ \text{Cost} &= 44 \times 1.4 = 61.6\ \text{p} \approx 62\ \text{p per day} \end{aligned}$$

Example 9.18

Calculate the total running cost of the following items of equipment if electricity costs 1.4 p per unit (kW h): (a) four 100-W lamps for 15 h, (b) one 250-W television receiver for 4 h, (c) one 750-W electric iron for 45 min, (d) one 15-W soldering iron for $2\frac{1}{2}$ h.

Solution

$$\text{(a) Cost} = 4 \times \frac{100}{1000}\ \text{kW} \times 15\ \text{h} \times 1.4\ \text{p/kW h} = 8.4\ \text{p}$$

$$\text{(b) Cost} = \frac{250}{1000} \times 4 \times 1.4 = 1.4\ \text{p}$$

$$\text{(c) Cost} = \frac{750}{1000} \times \frac{45}{60} \times 1.4 = 0.7875\ \text{p}$$

$$\text{(d) Cost} = \frac{15}{1000} \times 2.5 \times 1.4 = 0.0525\ \text{p}$$

$$\text{Total cost} = 10.64\ \text{p}$$

9.6 ELECTRICAL MEASUREMENTS

For various purposes it is necessary to make measurements to determine the magnitude of different electrical quantities. The most important of these are *current* and *potential difference*. If these two values are known it is possible to derive *resistance* and *power* by calculation.

9.6.1 The Moving-coil Meter

A description of the working principle of this widely used device is given in section 10.4.1. The meter is really a small electric motor whose movement is restrained by two helical springs, so that the deflection of the pointer is proportional to the magnitude of the *current* flowing. The springs cause the pointer to oscillate either side of the final position, which is undesirable since observations would be inaccurate and slow. *Damping* is therefore provided, by the use of a separate aluminium former for the coil. When this former (or frame) moves in the magnetic field it acts as a short

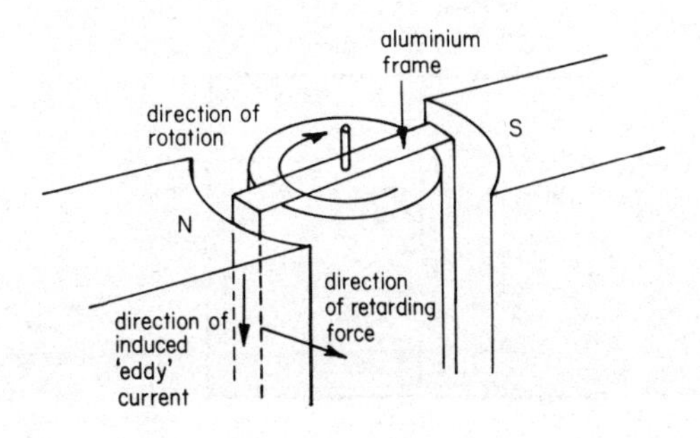

Figure 9.20

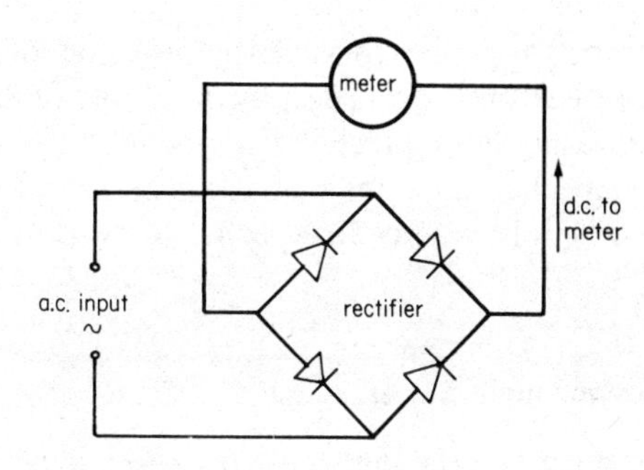

Figure 9.21

circuited generator winding (see figure 9.20). From the theory of electromagnetic induction, the currents induced in the sides of the frame as it moves across the field produce a force opposing the direction of motion of the frame (section 10.5.2). Consequently the movement is quickly arrested. These currents are known as '*eddy*' *currents* because they only circulate within the aluminium frame.

The current is supplied to the moving coil via the two helical springs already mentioned. This technique avoids problems due to flexible leads or slip rings which would interfere with the free movement essential for accuracy.

The magnetic field is uniform over the whole range of movement so the scale of this type of meter is equally divided (a proportional or evenly divided scale). The deflection of a moving-coil meter depends on the direction of current flow and incorrect connections cause the pointer to move backwards until it hits the 'stop' provided. For this reason also, the simple meter cannot measure alternating currents (a.c.), the deflections produced having a mean value of zero. In some applications a rectifier is connected in the meter circuit so that a.c. may be measured successfully, although the meter is handling current in only one direction (figure 9.21).

9.6.2 The Moving-iron Meter

This meter operates on a quite different principle from the moving-coil instrument. Two small pieces of soft iron are located in the central space of a coil of wire having many turns (figure 9.22).

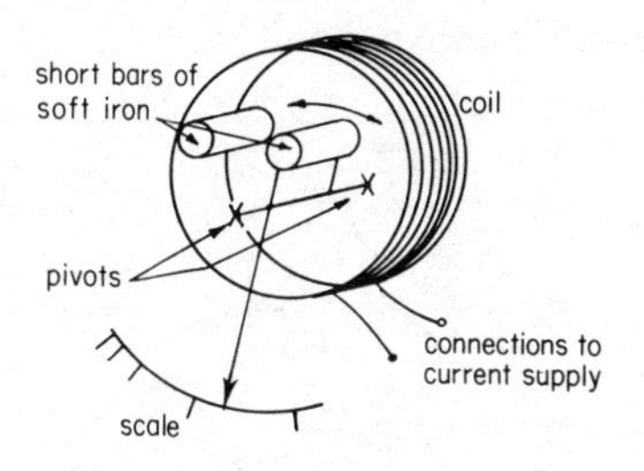

Figure 9.22 Moving-iron meter

The current being measured flows through this coil setting up a magnetic field along its axis which magnetises the two pieces of iron. The polarities of these temporary magnets are both in the same sense, that is, the like poles are adjacent, so a repulsive force is set up between them (section 10.1.2). One piece of iron is fixed to the coil frame but the other is attached to a pivoted arm and is able to move under the action of the repulsive force. A control spring limits the movement, as in the moving-coil meter, and damping is provided mechanically, usually by an air dashpot or vane.

Unfortunately the repulsive force depends on an inverse square law relationship with distance, that is, if the separating distance is doubled the force is reduced to one-quarter of its original value. This means that the divisions on the meter scale are unequal, becoming cramped together towards the position of maximum deflection. This disadvantage is offset by the ability of the moving-iron meter to measure both direct and alternating currents without modification, since the similar magnetisation of the two iron pieces always creates a repulsive force irrespective of the direction of the current.

9.6.3 The Thermocouple Meter

This device makes use of the *thermoelectric effect*, in which a small current is produced when the junction between two dissimilar metals is heated (figure 9.23). The current to be measured is passed through a small heating element which raises the temperature of such a junction and the resulting thermoelectric current is measured by a sensitive moving-coil meter.

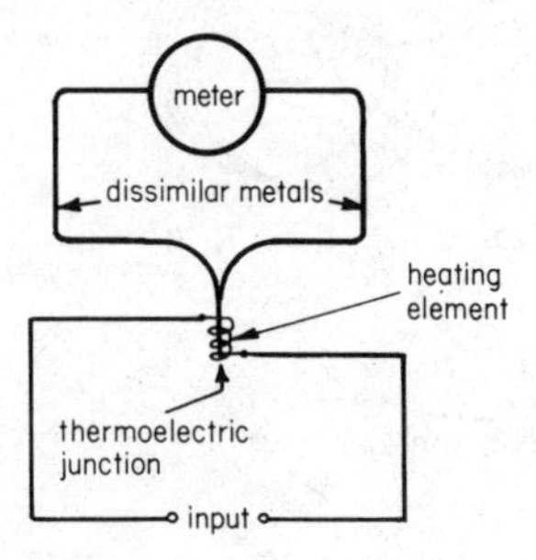

Figure 9.23 Thermocouple meter

9.6.4 Voltmeters and Ammeters

All the meters described above can be used to give both voltage and current readings.

The Voltmeter

This is *always connected in parallel* with the circuit components since it indicates the potential difference (or pressure difference) between two parts of the circuit. Ideally it should consume zero current.

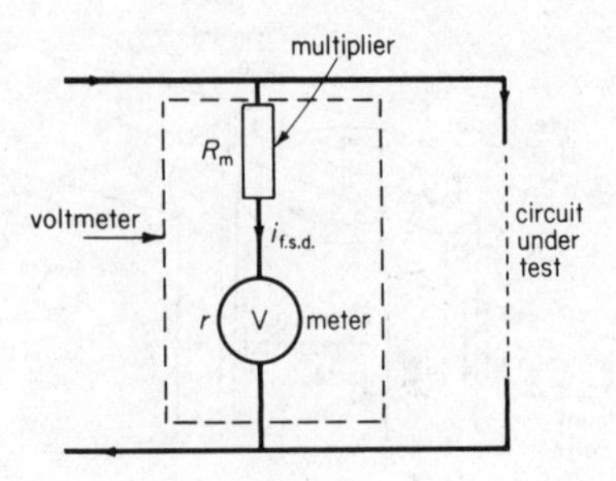

Figure 9.24 Use of a multimeter

To avoid the meter acting as a short circuit, a high resistance called a *multiplier* is connected in series with it (figure 9.24). Ideally the voltmeter should have an infinitely large resistance so it takes zero current from the circuit. This is impossible for a moving-coil meter since a finite current, however small, is needed to operate the meter. (The ideal voltmeter is an electronic device such as the cathode-ray oscilloscope which does actually use zero current (section 9.6.5).) To determine the value of multiplier required we need to know the value of the meter resistance r. Let the required reading of the meter be V volts. The meter current required for *full-scale deflection* will be $i_{\text{f.s.d.}}$ A. By Ohm's law

$$V = i_{\text{f.s.d.}}(R_m + r)$$

Hence the multiplier resistance is

$$R_m = \frac{V}{i_{\text{f.s.d.}}} - r$$

(The values of $i_{\text{f.s.d.}}$ and r for the meter are often marked on the dial face.) By selection of suitable multiplier values the meter can be adapted to any required range of voltages.

Example 9.19

A moving-coil meter of resistance 50 Ω and f.s.d. current 1 mA is to

rough which observed cur-
voltage is developed is given

$\left.\frac{V}{I}\right)$

$\times 100 = 0.75\,\%$

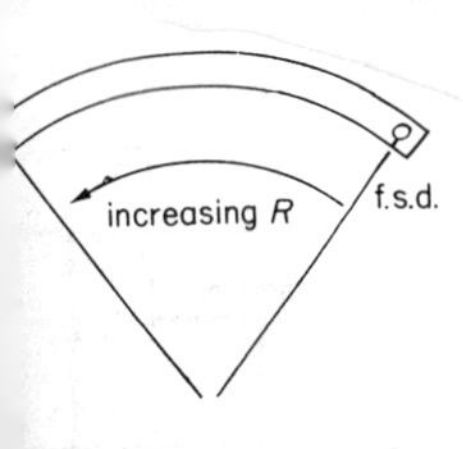

eter and ammeter, the value
by an instrument called an
resistance R is connected in
1.5 V), a milliammeter and a
and B are connected directly
ive full-scale deflection of the
ointer is calibrated for *zero*
values of resistance are then
produce smaller deflections
scale. The meter is dependent
by the cell, and S is kept large
void errors due to its internal
ro the meter each time before
ore precise methods available
and convenient for checking
ccuracy, and the results are

But $V = i_{\text{f.s.d.}} \times r$, hence

$$R_s = \frac{i_{\text{f.s.d.}}}{I - i_{\text{f.s.d.}}} \times r$$

Example 9.20

The meter used in example 9.19 is now to be used as an ammeter with a maximum reading of 5 A. Calculate the value of shunt to be used.

Solution

$$R_s = \frac{i_{\text{f.s.d.}}}{I - i_{\text{f.s.d.}}} \times r$$

$$= \frac{1 \times 10^{-3}}{5 - 1 \times 10^{-3}} \times 50 = 0.010002 \approx 0.01\,\Omega$$

(The error involved when using the approximate value is very small, 0.02 per cent.)

The Ammeter/Voltmeter Method for Measuring Resistance

From Ohm's law the resistance is given by

$$R = \frac{V}{I}$$

If we can measure the voltage drop (or potential difference) across a resistance and the current flowing through it, the resistance can be obtained (figure 9.26).

Errors Involved If we take into account the effect of the meters on the circuit, we should note that not all the current passes through R; some must operate the voltmeter. Provided R values are small compared with the resistance of the voltmeter this method is acceptable and current errors are small. For R values approaching the resistance of the voltmeter, the ammeter should be connected in series with R as shown in figure 9.27. This ensures that the current

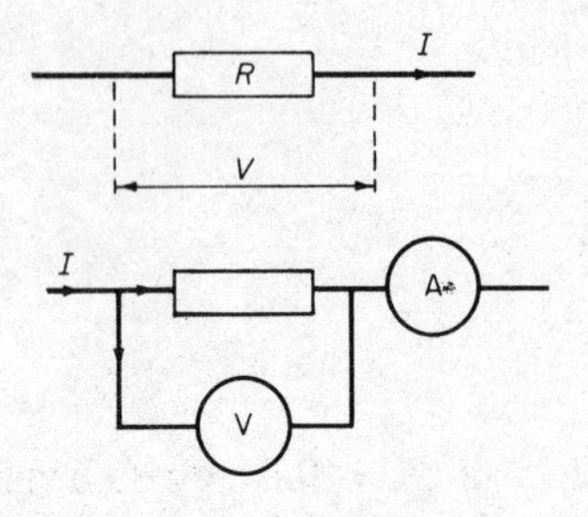

Figure 9.26

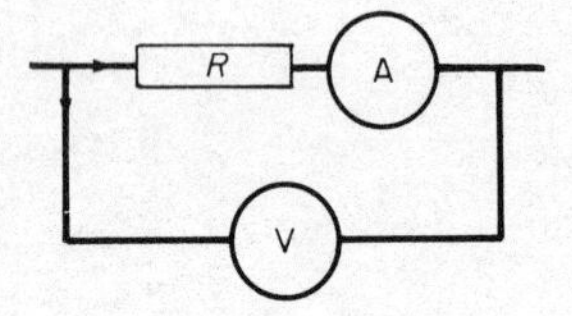

Figure 9.27

indicated is the true current and the voltage drop across the ammeter is very small compared with V, producing a very small error. *Note* the method always produces a small error.

Example 9.21

An ammeter of resistance 0.5 Ω and a voltmeter of resistance 2000 Ω are used to measure a resistance of 15 Ω. Calculate the percentage error involved in this method. The supply is a steady 2 V.

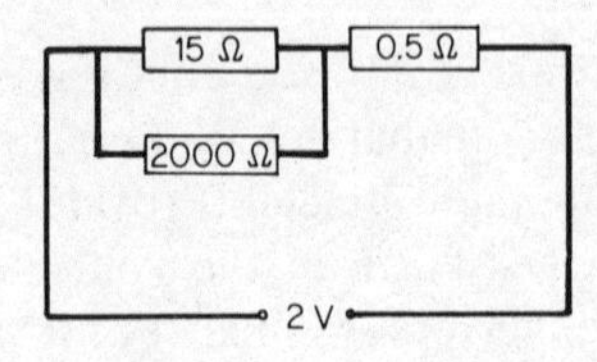

Figure 9.28

Solution The apparent resistance
rent flows and across which observe
by

$$R = \frac{2000 \times 15}{2000 + 15} = 14.888\ \Omega \quad ($$

$$\text{percentage error} = \frac{(15 - 14.888}{15}$$

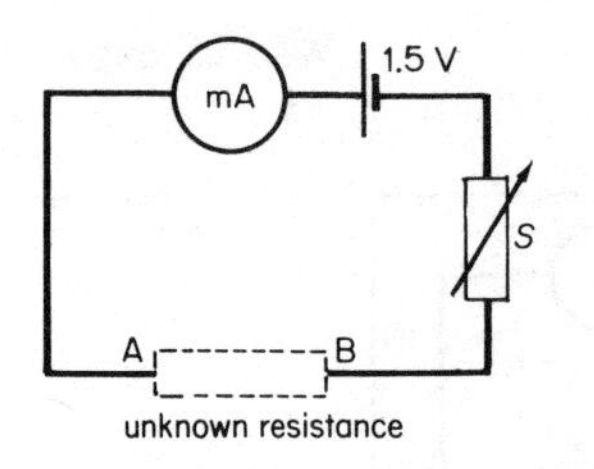

Figure 9.29 The ohmmeter

The Ohmmeter

Apart from the technique using volt
of a resistance may be measured
ohmmeter (figure 9.29). The unknow
series with a small cell or battery (sa
variable limiting resistor S. When A
together the value of S is adjusted to
milliammeter. This position of the
resistance. If various other standar
connected across AB, the meter wi
which can be suitably marked on th
on a steady voltage being maintaine
to minimise the drain on the cell and
resistance. Even so it is necessary to
use, by adjustment of S. There are n
than the ohmmeter, but it is quick
values within reasonable limits of

indicated directly without the need for calculations. From Ohm's law, $I = V/R$. (V is constant.) Therefore the current is inversely proportional to the value of R and the scale will be unequally divided, with large values cramped together at the beginning of the scale.

The Multimeter

A most useful piece of equipment is the multimeter, which consists of a sensitive moving-coil meter, able to operate with very low current consumption. A large number of shunts and multipliers is built into the case, to be selected as required by rotary switches, enabling the meter to be used as an ammeter or voltmeter over a very wide range. In addition a cell and associated circuitry are often provided to permit use of the multimeter as an ohmmeter.

A universal multimeter also has a rectifier which can be switched into circuit for a.c. measurements.

9.6.5 The Cathode-ray Oscilloscope

The cathode-ray oscilloscope (c.r.o.) is operated by *electrostatic* rather than electromagnetic effects, to measure voltages. Instead of mechanical components which show the measurements as deflections of a pointer or needle, it uses the lightest and most sensitive indicator in existence—a beam of electrons. The oscilloscope consists essentially of an evacuated glass envelope containing three main parts

(1) *the electron gun* (figure 9.30): This produces a finely focused beam of electrons which travels in a straight line from the outlet aperture of the gun;

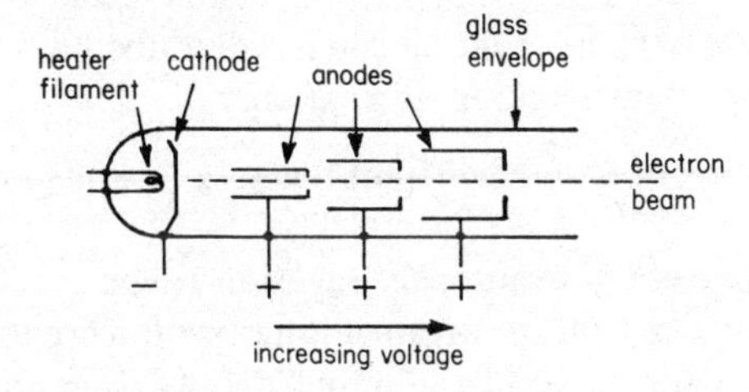

Figure 9.30 The electron gun

(2) *the deflector plates*: these are used to deflect the electron beam from its normal straight path. One plate is positively charged and the other negatively charged, using the voltage to be tested. The electrons, which themselves have a negative charge, are deflected by the electric field between the plates; the beam is attracted by the positive plate and repelled by the negative plate; the ultimate deflection is proportional to the voltage applied to the plates (figure 9.31);

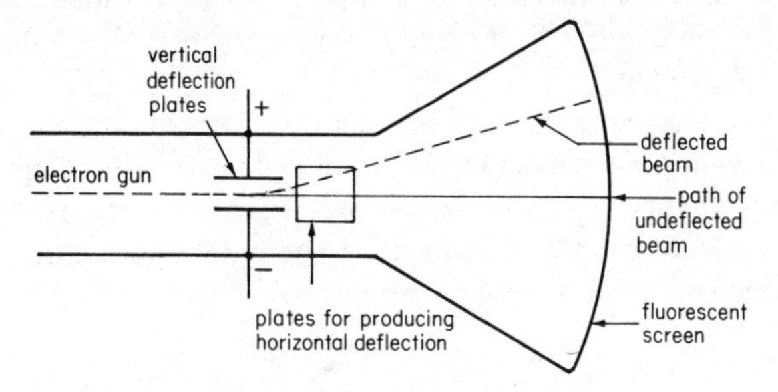

Figure 9.31 Beam deflection

(3) *the fluorescent screen*: the electron beam, which is invisible to the naked eye, is directed on to a screen of fluorescent material which glows brightly at that point, producing a spot of light. The deflection of the spot can then be used to give the value of the voltage applied to the plates (figure 9.32).

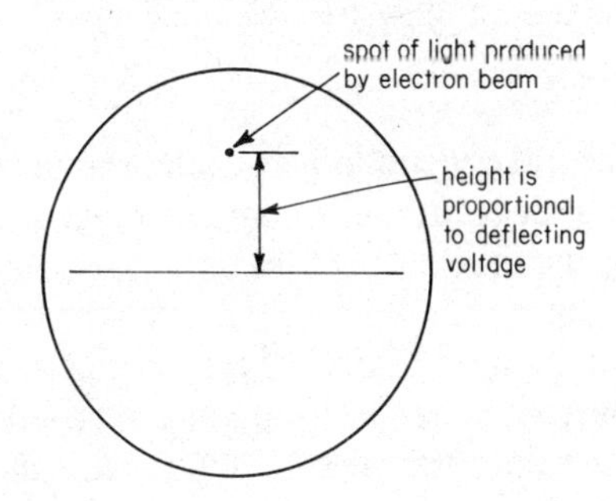

Figure 9.32 Oscilloscope screen

Advantages of the Oscilloscope

Although the oscilloscope is more complicated and expensive than ordinary meters, it has advantages which make it an extremely useful piece of equipment. It does not consume any current from the circuit under test, so the circuit voltages are the *true values* and are not modified by the need to supply a meter current (as happens with moving-coil meters).

The oscilloscope can be used to measure current if a resistance of known value is connected in the circuit and the voltmeter used to measure the p.d. between its ends. The current is then calculated using Ohm's law.

The oscilloscope can also be used to display the change of voltage *with time*, particularly helpful in the study of alternating-current circuits. A second set of plates is provided, arranged at right-angles to the first set and able to deflect the spot horizontally (figure 9.33).

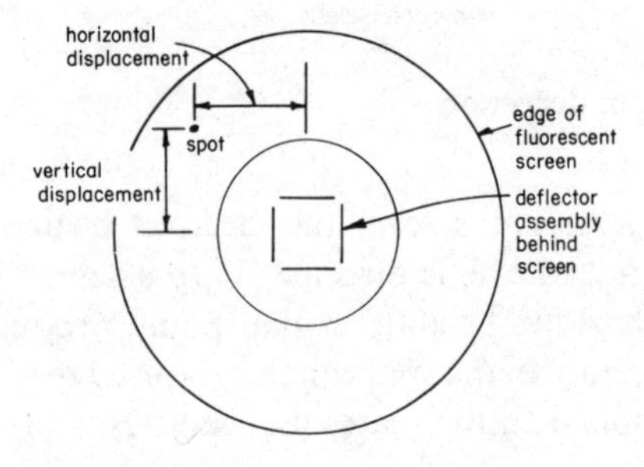

Figure 9.33 Displacement of spot in two dimensions

Note The vertical deflection uses plates arranged horizontally and the horizontal deflection uses plates arranged vertically, so to avoid confusion they are referred to as X and Y plates (as for a graph)—figure 9.34.

A special circuit in the oscilloscope can provide a varying voltage which deflects the spot *horizontally* at a steady speed (known as the timebase). Owing to persistence of vision the moving spot forms a continuous line or *trace*, which is in effect a graph of voltage plotted against time (figure 9.35).

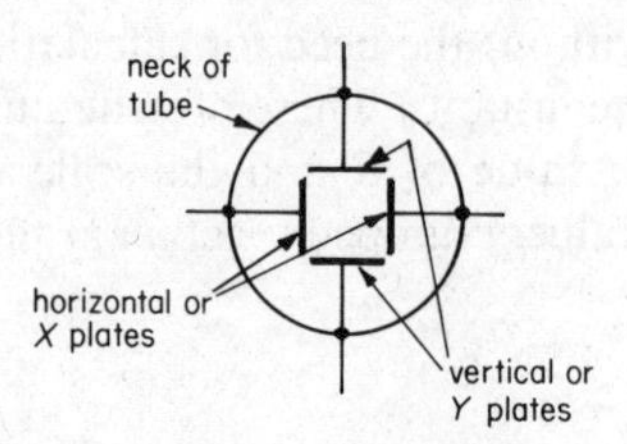

Figure 9.34 Arrangement of X and Y plates

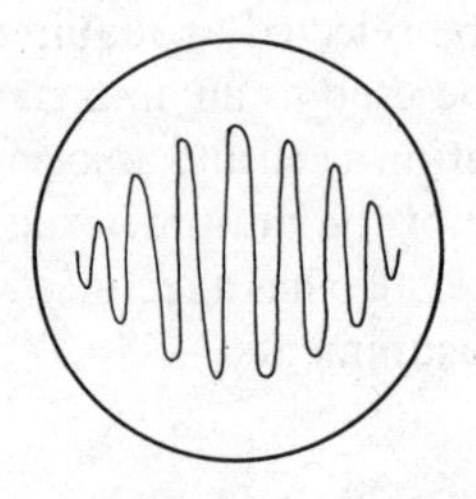

Figure 9.35 Typical complex waveform using X and Y plates together

When the spot reaches the edge of the screen it is made to return to the beginning of its 'sweep' rapidly, so that it leaves no trace and only the forward sweep is displayed. The timebase allows variations of voltage to be observed continuously.

9.7 MEASUREMENT IN A.C. SYSTEMS

The generation of an alternating e.m.f. is described in section 10.6. The most important features to note are

(1) the direction of the current reverses completely at regular intervals

(2) the current continuously varies in value.

The graph of e.m.f. plotted against time for a full cycle is as shown in figure 9.36. The value of the induced e.m.f. E (section 10.6) is related to the position angle of the generator coil θ, thus

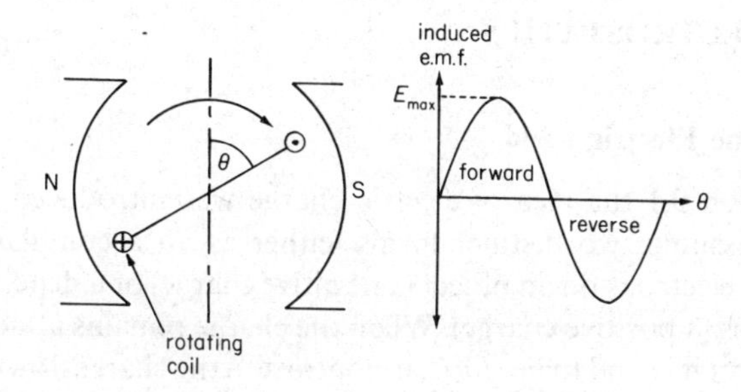

Figure 9.36 Alternating e.m.f.

$$E = E_{max} \sin \theta$$

From which a similar equation is obtained for current

$$I = I_{max} \sin \theta$$

The values of E_{max} and I_{max} are called *peak values* of e.m.f. and current, but these cannot be used as *effective values* since they exist for only an instant.

9.7.1 Average or Mean Value

One possibility is to take the *average value* of the current or e.m.f. (also called the *mean value*), but over a complete cycle this is zero and meaningless, since we *know* that current has been flowing. Instead we consider half a cycle, and obtain the average value for that period. It can then be shown that the mean value is given by the following formulae

$$E_{mean} = E_{max} \times 0.6366$$

or

$$I_{mean} = I_{max} \times 0.6366$$

(*Note* Using the factor 0.6366 is only valid when the waveform is *sinusoidal*.)

In section 9.5 it was shown that the energy converted into heat is proportional to the *power* dissipated in a conductor, and that power is proportional to the *square of the current flowing*. If we wish to calculate the heating effect of an alternating current we should strictly have to measure every possible instantaneous value of current and then square each one to find the power at that instant. (This, of course, is a difficult task, because theoretically there is an infinite number of instantaneous values.) From these we could then find the *mean square* value of the current. But what we really wish to know is the equivalent *steady* current which produces the *same heating effect* as the alternating current. Therefore we must obtain the *square root* of the mean square of the current. This is termed the *root mean square* or *r.m.s.* current. Using an appropriate mathematical method it can be shown that, for a sinusoidal waveform, the r.m.s. value is given by

$$\text{r.m.s. value} = \frac{\text{peak value}}{\sqrt{2}}$$

$$= \text{peak value} \times 0.707$$

Example 9.22

Calculate the r.m.s. value of a sinusoidal alternating current if the peak value is 0.25 A.
Solution

$$\text{r.m.s. value} = \frac{0.25}{1.414} = 0.1768 \text{ A}$$

Example 9.23

Calculate the peak value attained by a sinusoidal a.c. voltage of r.m.s. value 240 V.
Solution

$$\text{Peak value} = \text{r.m.s. value} \times \sqrt{2}$$
$$= 240 \times 1.414$$
$$= 339 \text{ V}$$

We thus have three different methods of representing the magnitude of an alternating value

(1) peak value (the maximum in one direction)
(2) mean value (peak value × 0.6366)
(3) r.m.s. value (peak value × 0.7071).

The most important of these is the r.m.s. value, which is the *effective* value of an alternating current. Most meters are calibrated to show r.m.s. values and therefore assume the existence of a sinusoidal waveform. *Power* dissipated by a resistance is obtained using the r.m.s. current or voltage.

Example 9.24

Calculate the power dissipated by a 60 Ω resistor which is connected to an a.c. supply with peak voltage 350 V.
Solution

$$\text{r.m.s. voltage} = \frac{\text{peak voltage}}{\sqrt{2}}$$

$$= \frac{350}{1.414}$$

$$= 247\ \text{V}$$

$$\text{power dissipated} = \frac{(V_{\text{r.m.s.}})^2}{R}$$

$$= \frac{247 \times 247}{60}$$

$$= 1017\ \text{W}$$

9.8 ELECTROSTATICS

9.8.1 The Electric Field

In section 9.1 the idea of electric charge was introduced, which could assume two distinct forms: either as an accumulation of surplus electrons on an object (a negative charge) or a deficiency of electrons (a positive charge). When the charge remains attached to the object it is said to be *static*, in contrast to the charge flowing in a conductor which is *current*. If two bodies with the same type of electrostatic charge (that is, both positive or both negative) are brought close together, a repulsive force is detected between them. On the other hand, two charges of opposite types (one positive, one negative) exert an attractive force on each other. In general: *like charges repel, unlike charges attract.*

It has been shown by careful measurements that the magnitude of the force is directly proportional to the product of the two charges and inversely proportional to the square of the distance separating them. Thus

$$F \propto \frac{q_1 q_2}{r^2}$$

$$F = \frac{q_1 q_2}{r^2} \times \text{a constant}$$

This is called *Coulomb's law*. The forces are very large. If the charges are each 1 coulomb and the separation is 100 m, the force value is about 1000 kN (the gravitational force on a mass of 100 t). At 1 m separation the force is ten thousand times this value. Usually static charges are very much smaller than 1 coulomb and the forces are more manageable.

The zone of influence surrounding a charge is called an *electric field* (or electrostatic field). It is a vector quantity, having both magnitude and direction, and is defined in terms of the force exerted on a test charge q placed in the field. For a given location the electrostatic force F_e on q is given by

$$F_e \propto q$$

$F_e = q \times$ a constant, E, for that location

field strength, $E = \dfrac{F_e}{q}$ N/C

9.8.2 The Capacitor

A capacitor is a simple device for storing a charge. It consists of two parallel plates made of thin foil or metal sheet, separated by an insulating material called a *dielectric*. If the plates are connected to either side of a steady source such as a battery, a positive charge will accumulate on one plate and an equal negative charge on the other (figure 9.37). Thus a potential difference will be developed across the dielectric which must be a good enough insulator to maintain the p.d. without breaking down; also a means must be provided to resist the forces which pull the two plates together.

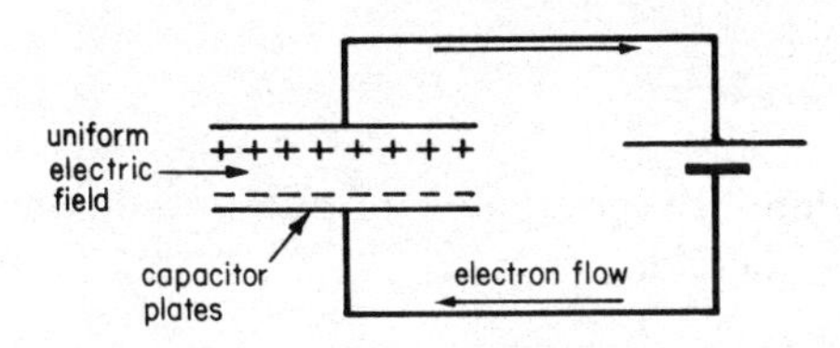

Figure 9.37 Accumulation of charge in a capacitor

It can be shown that the electric field between the plates is *uniform* and that the *potential gradient* in the field is also uniform (expressed in V/m), that is, the potential changes uniformly as we move from one plate to the other (assuming the dielectric remains the same throughout). The potential gradient is also numerically equal to the magnitude of the electric field strength.

The ability to store a charge is called *capacitance* (symbol C).

Definition of Capacitance

If the potential V across the plates of a capacitor is *one volt* when it is storing a charge Q of *one coulomb* the capacitance C is said to be *one farad* (F).[7]

The relationship is given by the equation

$$Q \text{ (coulombs)} = V\text{(volts)} \times C\text{(farads)}$$

The farad is a very large unit and capacitors in common use are usually of capacitance of the order of microfarads (μF, 10^{-6}F) or even picofarads (pF, 10^{-12}F).

Example 9.25

Calculate the charge stored when a source of 50 V is connected across a 40 pF capacitor.

Solution

$$\begin{aligned} Q &= VC \\ &= 50 \times 40 \times 10^{-12} \\ &= 2 \times 10^{-9} \text{ C} \end{aligned}$$

Experiments show that the value of the capacitance depends on three quantities

(1) area of the plates, A
(2) distance separating the plates, d
(3) electrical properties of the dielectric (called the *permittivity*), ε

$$C \propto A$$

$$C \propto \frac{1}{d}$$

$$C \propto \varepsilon$$

$$C = \varepsilon \frac{A}{d}$$

Note on Permittivity The property of the insulator ε, known as permittivity, is a property of matter which varies according to the nature of the substance used as a dielectric. It relates the charge distribution on the plates of the capacitor σ (coulombs/m^2) to the electric field E between them. (Note, σ is here used for a different purpose from that in section 9.4.1.)

$$\varepsilon = \frac{\sigma}{E}$$

For free space (a vacuum) the value is 8.85×10^{-12} (symbol ε_0) and has units of *farads per metre*. Other materials are usually specified in terms of their *relative permittivity* from the formula

$$\text{relative permittivity, } \varepsilon_r = \frac{\text{absolute permittivity, } \varepsilon}{\text{permittivity of free space, } \varepsilon_0}$$

Hence by definition ε_r for a vacuum = 1.

Example 9.26

Calculate the capacitance of a capacitor using plates of area 800 cm^2 and a dielectric made of polythene sheet 0.1 mm thick (ε_r = 2.3).

Solution

$$C = \varepsilon \frac{A}{d}$$

$$= \frac{(2.3 \times 8.85 \times 10^{-12}) \times (800 \times 10^{-4})}{0.1 \times 10^{-3}}$$

$$= 1.628 \times 10^{-8} \text{ F}$$
$$= 0.0163 \ \mu\text{F}$$

Energy Stored in a Capacitor

A capacitor maintains a charge Q coulombs at a potential V volts, therefore it must retain a store of energy in *joules* which can be recovered when the capacitor is discharged again.

Consider a capacitor whose charge is steadily increased from zero until it has a value Q coulombs. The charge–potential graph will appear as in figure 9.38.

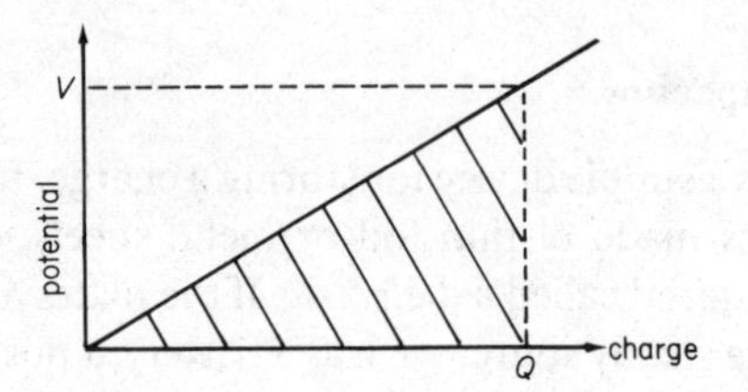

Figure 9.38

The energy stored in joules is the area under the graph

$$\text{area, } E = \frac{Q \times V}{2}$$

[charge × potential = coulombs × (joules/coulombs) = joules]
Now $Q = VC$, therefore

$$E = \frac{VC \times V}{2}$$

or

$$E = \tfrac{1}{2}CV^2 \text{ J}$$

Example 9.27

Calculate the energy stored in a 50-μF capacitor when charged to its full potential of 1000 V.

Solution

$$E = \tfrac{1}{2}CV^2$$
$$= 50 \times 10^{-6} \times 1000 \times 1000 \times \tfrac{1}{2}$$
$$= 25 \text{ J}$$

Use of Capacitors in A.C. Circuits

When a capacitor is first connected to a d.c. source a brief current flows which dies away as the full charge is reached and the p.d. across the plates approaches that of the supply voltage (figure 9.39). If the supply polarity is now reversed there will be a current in the opposite direction due to the discharge of the capacitor and its accumulation of a charge of opposite sign. This is shown in figures 9.40a and b. If the polarity is again reversed there will be a further reversal of the current in the circuit as the charge is restored to its original value. An *alternating e.m.f.* will produce forward and reverse currents in endless succession and we have the *apparent* effect of passing an alternating current *through* the capacitor.

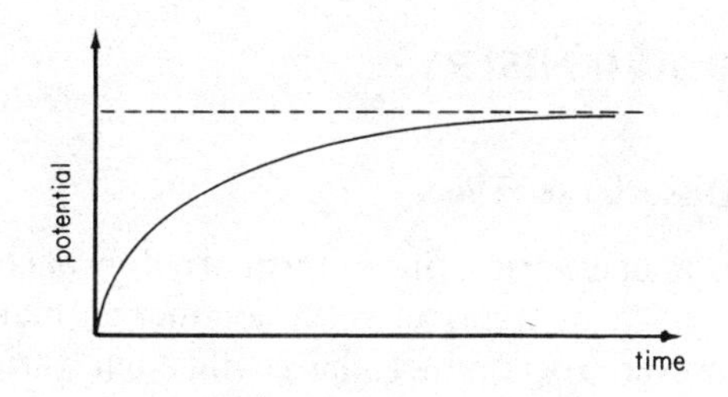

Figure 9.39

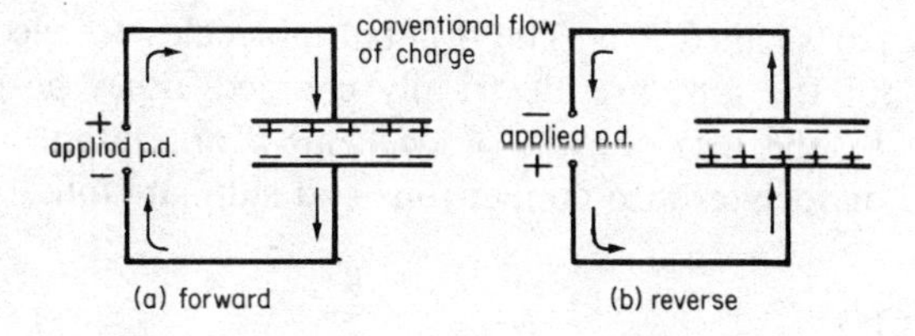

Figure 9.40 Alternating p.d. applied to a capacitor

Capacitor Construction

All capacitors use thin metal foil or metal films for the plates on which the charge is distributed, but constructional design varies considerably in the choice of dielectric, depending on the usage.

Plastics (figure 9.41) Two strips of foil are interleaved with two strips of a plastics such as polystyrene or polythene (ε_r between 2 and 4). The whole is then rolled into a cylinder with the connections to the two foil plates emerging at opposite ends. The complete unit is enclosed in a suitable protective case.

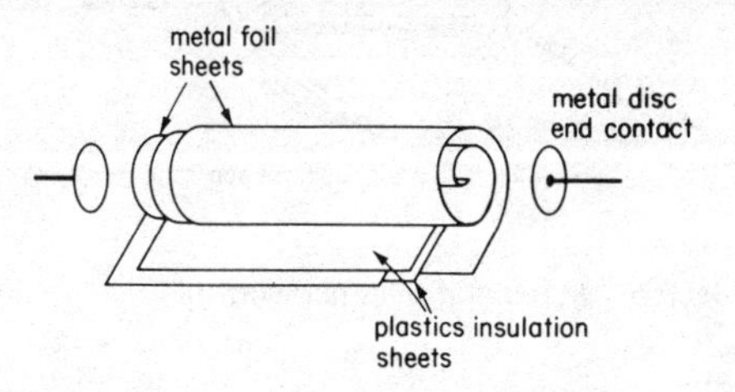

Figure 9.41 Plastics capacitor construction (without cover)

Mica (figure 9.42) Thin sheets of the mineral mica (ε_r value 7) are used, interleaved with foil. Mica is too brittle to roll in a cylinder so it is 'stacked' in a sandwich with alternate foils connected to the two connecting wires. A moulded plastics case is then formed around the capacitor. Sometimes a silver film is deposited on the mica instead of foil, with a consequent reduction in size.

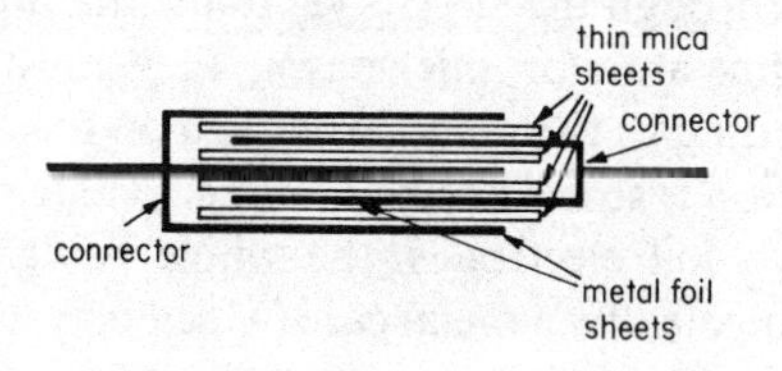

Figure 9.42 Mica insulated capacitor (without cover)

Ceramic Very good results can be obtained with certain ceramic materials which contain steatite (ε_r value 8) but other materials, notably barium titanate (ε_r value up to 1200), can also be used to produce capacitors of a *very small physical size.* The ceramic is often in the shape of a tube with the plates formed by silvered films deposited on the inside and outside surfaces.

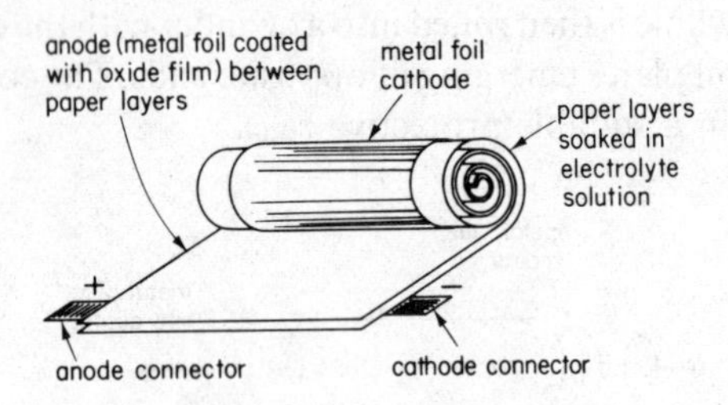

Figure 9.43 Electrolytic capacitor construction

Electrolytic (figure 9.43) These make use of the extremely thin (0.0001 mm) film of oxide formed on an aluminium plate which acts as the anode of an electrolytic cell containing a solution of ammonium borate (see section 9.9). The oxide film is an excellent insulator with a maximum potential gradient of 10^9 V/m. The foil carrying the film is one plate of the capacitor and the other plate is the *electrolyte itself*, with the connection made via the cathode of the cell. Since $C \propto 1/d$ the electrolytic method can produce *very large capacitances* particularly for low voltage work when d can be very small indeed. However, they do require a slight d.c. 'leakage' current to pass through the electrolyte to maintain the continuity of the oxide film, and for this reason they must be correctly connected according to the polarity marked on the terminals. The electrolyte solution is stored in absorbent paper strips sandwiched between the two foil electrodes, the whole being rolled into a cylinder and protected by a metal case (which may form one of the external connections).

Variable Capacitors

When it is necessary to use capacitors of variable capacitance value, the dielectric may be mica or plastics sheets or even air, though the latter requires a large separating distance between the plates.

One set of plates is fixed while the other is moveable, as shown in

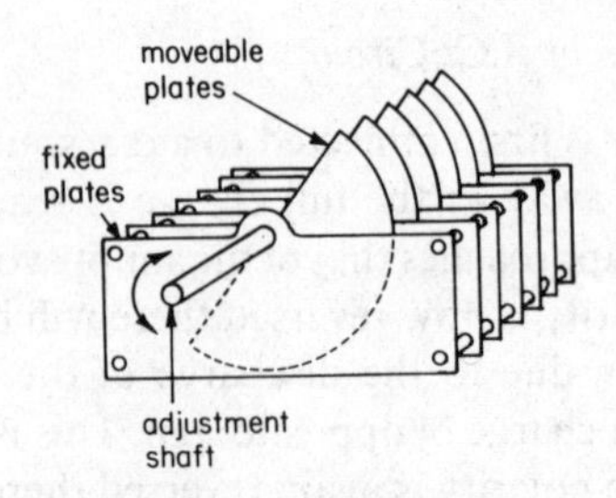

Figure 9.44 Variable capacitor (air dielectric)

figure 9.44, thus allowing the *effective area* of the two plates to be altered. This is widely used in the tuning circuits of radio receivers.

9.9 ELECTROCHEMISTRY

9.9.1 Ionic Dissociation Theory

The conduction of electricity in the form of a flow of charge has so far been discussed in terms of solid conducting materials. It is possible, however, to conduct charge through various liquids known as electrolytes. Unlike solids, these liquids use a different method of transporting the charge as it flows through them. Electrolytes are usually solutions of acids, bases (alkalis) or salts and as such contain molecules which can *dissociate* in solution into separate parts called *ions*. The original molecules are electrically neutral but the ions are electrically charged either positive or negative by the loss or gain of electrons. For example, copper sulphate dissociates into copper ions and sulphate ions.

$$CuSO_4 \rightarrow Cu^{2+} + SO_4{}^{2-}$$

Each sign indicates a charge equal to one electron, that is, Cu^{2+} means a copper atom with two positive electronic charges. Many other materials form ions in a similar way

$$ZnSO_4 \rightarrow Zn^{2+} + SO_4{}^{2-} \text{ (zinc sulphate)}$$
$$NaCl \rightarrow Na^{+} + Cl^{-} \text{ (common salt)}$$

$$H_2SO_4 \rightarrow 2H^+ + SO_4^{2-} \text{ (sulphuric acid)}$$
$$HCl \rightarrow H^+ + Cl^- \text{ (hydrochloric acid)}$$

and so on. The charges are always equal so that the solution remains *electrically* neutral at all times.

9.9.2 Electrolysis

If a pair of electrodes is immersed in an electrolyte and a potential difference is applied to them from a direct current source, the ions are attracted by the electric field between unlike charges and *migrate* towards the appropriate plates (figure 9.45). The assembly is called an *electrolytic cell* and the process is called *electrolysis.*

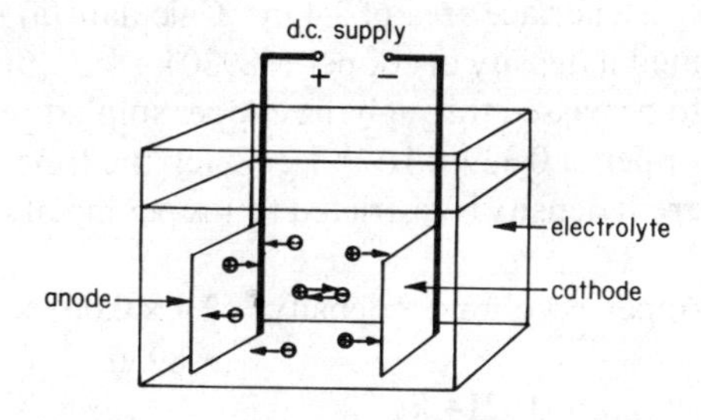

Figure 9.45 Migration of ions in electrolytic cell

The ions that carry a positive charge migrate to the negative plate or *cathode.* On arrival they acquire additional electrons from the external source of supply and cease to be ions, changing back into neutral atoms of the elements involved. Similarly the negatively charged ions migrate to the positive plate or *anode* where they give up their surplus electrons and also become neutral atoms. These electrons then form the current flowing in the external circuit.

Consider the process for the electrolysis of sodium chloride. At the anode

$$Cl^- \rightarrow Cl + e^-$$

(e represents an electron). At the cathode

$$Na^+ + e^- \rightarrow Na$$

Note that the formation of the elements sodium and chlorine has occurred in *different locations* and they can be collected separately if required, so this technique can be used to separate salts and other electrolytes into their constituent elements. The process is widely used in chemical manufacture.

The electrolytic cell is sometimes called a voltameter—*not* a voltmeter—(after Alessandro Volta) and a simple experiment uses a Hofmann voltameter to split up water into hydrogen and oxygen by electrolysis (figure 9.46).

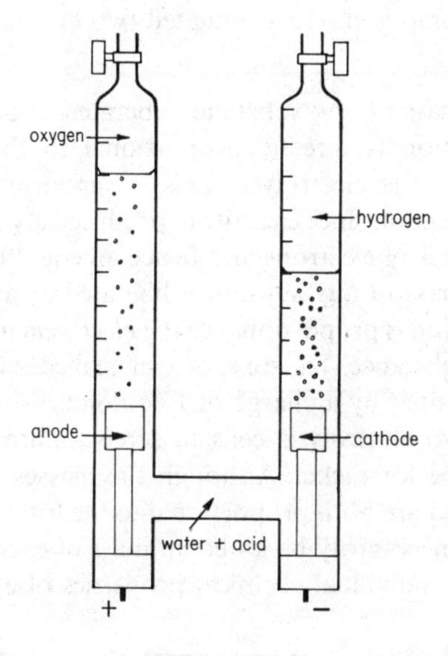

Figure 9.46 Hofmann voltameter for electrolysis of water

9.9.3 Electroplating

If we use a solution of blue copper sulphate as the electrolyte, we find that continuous passage of current makes the solution become paler in colour until all the copper sulphate is used up. To avoid this we can make use of a copper anode, so that the liberated sulphate groups formed on its surface combine with the copper to form *fresh copper sulphate* which then dissolves and prevents the solution becoming any weaker. At the same time the copper liberated at the cathode is deposited as a thin layer over the surface. This is the basis of *electroplating*. Copper plating is carried out as described above; nickel plating is similar but uses nickel sulphate solution. Many other metals, for example, chromium, can be deposited from their salts in the same way.

9.9.4 Faraday's Laws of Electrolysis

Faraday investigated the behaviour of electrolytic cells and attempted to define the factors which affected the amount of substance decomposed. He formulated two *laws of electrolysis* as follows.

(1) The mass of any substance liberated at an electrode by electrolytic action is directly proportional to the *total charge* passed through the electrolyte. This is supported by the ionic theory of conduction since each atom produced always involves an identical number of electrons and hence an equal electric charge.

(2) The mass of any substance liberated at an electrode by electrolytic action is proportional to the *electrochemical equivalent* (e.c.e.) of the substance. The e.c.e. of a substance is the mass of the substance liberated by a charge of 1 coulomb.

Consider two electrolytic cells in series (figure 9.47), one for copper and one for nickel. Although the masses of copper and nickel deposited are both proportional to the total charge passed, there will not necessarily be an equal mass of each metal. It will depend on the individual chemical properties of each metal.

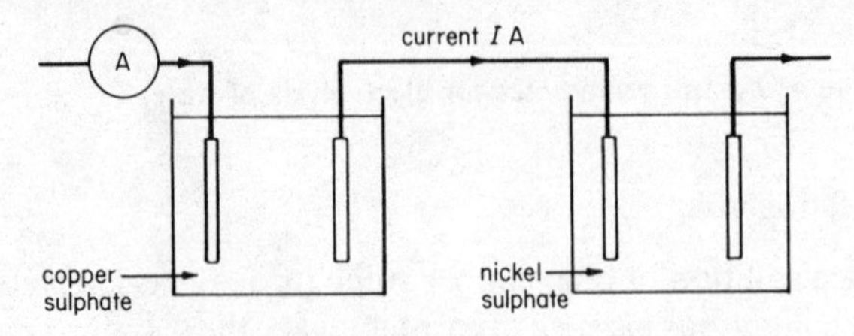

Figure 9.47 Electrolytic cells in series

The two laws are both expressed in the equation

$$m = ZQ$$

where Z is the e.c.e. of the substance and Q is the charge in coulombs. Alternatively

$$m = ZIt$$

where I is the current in amperes flowing for time t.

Example 9.28

It is required to deposit a layer of copper 0.001 mm thick on a component with a surface area of 2.4 m^2. Calculate (a) the mass of copper deposited if density of copper is 8930 kg/m^3, (b) the charge in coulombs to be passed through the copper sulphate electrolyte if the e.c.e. of copper is 0.329×10^{-6} kg/C, (c) the time taken if the maximum current density is restricted to 1 A per m^2 of surface area.

Solution

(a) Mass of copper = volume × density = $2.4 \times 0.001 \times 10^{-3} \times 8930$

= 0.0214 kg

(b) $$\text{Charge in coulombs} = \frac{\text{mass}}{\text{e.c.e.}} = \frac{0.0214}{0.329 \times 10^{-6}}$$

= 65045 C

(c) Current = current density × surface area
= $1.0 \times 2.4 = 2.4$ A

$$\text{Time taken} = \frac{\text{charge}}{\text{current}} = \frac{65045}{2.4}$$

= 27102 s
= 7h 31 min 42 s ($\approx 7\frac{1}{2}$h)

9.9.5 The Voltaic Cell

When a cell uses two *different* metals for the anode and cathode,

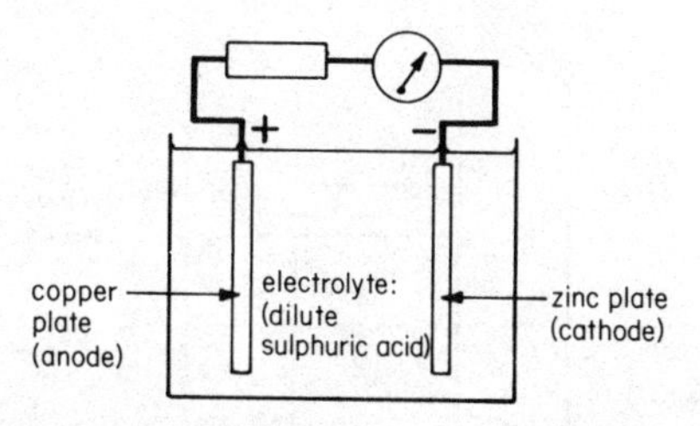

Figure 9.48 Simple voltaic cell

together with a suitable electrolyte, an e.m.f. can be detected between the plates which will drive a current through an external circuit. This is known as a *voltaic cell* and a typical example is the simple cell using zinc and copper plates with dilute sulphuric acid as electrolyte (figure 9.48). Close observation of the working cell reveals the presence of bubbles of hydrogen on *both* plates but the formation of bubbles on the copper plate ceases when the external circuit is interrupted. We can think of the reactions as a form of electrolysis but without an *external* source of current. The electrolyte molecules dissociate forming hydrogen ions and sulphate ions

$$H_2SO_4 \rightarrow 2H^+ + SO_4{}^{2-}$$

At the copper plate the hydrogen ions form hydrogen atoms by acquiring electrons *from the external circuit*

$$2H^+ + 2e^- \rightarrow 2H = H_2 \text{ (molecule)}$$

At the zinc plate the sulphate ions give up their surplus electrons *to the external circuit* and then react with the zinc to form zinc sulphate which passes into solution.

Thus the external circuit is supplied with electrons via the zinc plate and loses them again via the copper plate: a current is created in the external circuit. The e.m.f. driving this current is provided by the chemical effect of the electrolyte on the zinc as it passes into solution, both the zinc and the acid being consumed by the reaction. This is an energy-releasing process which provides power for consumption in the external circuit.

The value of the e.m.f. depends only on the nature of the chemical reactions involved and is unaffected by the size of either the cell or its components.

There are two disadvantages which make the simple cell unsuitable for use as a source of current for long periods.

Polarisation The copper plate is a source of hydrogen bubbles during the operation of the cell and these blanket the surface, reducing contact with the electrolyte and lowering the output current. This effect is known as *polarisation.* The bubbles are removed by maintaining an *oxidising agent* round the plate to convert the hydrogen to water. In the Daniell cell the agent is copper sulphate solution kept separate from the acid electrolyte by a porous pot. The output is then unaffected by polarisation making the cell a useful and stable source for laboratory purposes (figure 9.49).

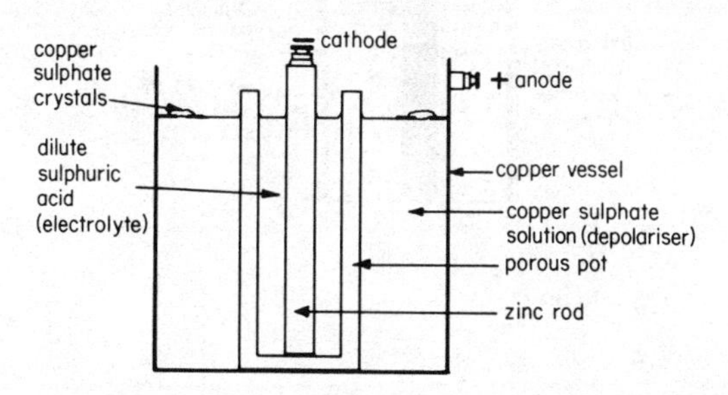

Figure 9.49 Daniell cell

Local Action The bubbles of hydrogen produced at the zinc plate are caused by zinc dissolving into the acid. This is a reaction which is independent of the cell function and is called *local action.* It results in a very short life for both the zinc and the acid but can be minimised by the application of a little mercury to the surface of the zinc. The mercury forms an alloy with zinc called an *amalgam,* which reduces local action but still allows the normal cell function to take place. The useful life of the zinc is thus greatly extended.

9.9.6 The Dry Cell (or Leclanché Cell)[8]

For a compact, portable supply of electricity the dry cell is used (figure 9.50). It consists of a zinc canister which acts as the cathode or negative terminal, housing a central carbon rod acting as the anode or positive terminal. The oxidising agent for depolarising the anode is a mixture of powdered manganese dioxide and carbon, which for convenience is contained in a cotton bag. The whole is saturated with a paste or jelly of ammonium chloride which acts as the 'dry' electrolyte. The e.m.f. is 1.5 V but the terminal voltage may be less due to polarisation during heavy use. A steel case is often provided for protection and to contain the electrolyte if the zinc canister is perforated by local action.

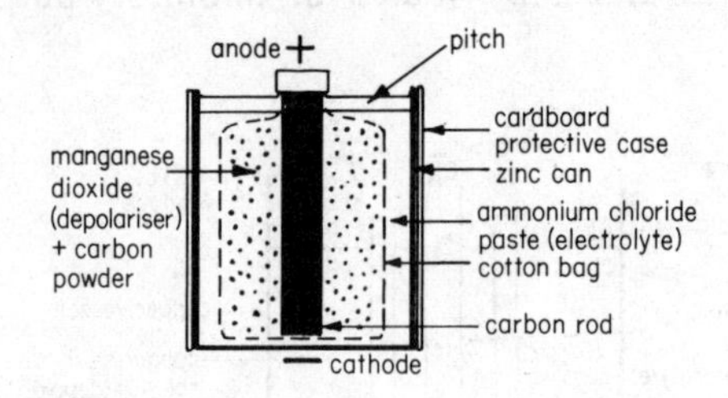

Figure 9.50 Leclanché cell

Dry cells are often referred to loosely as 'batteries' but the term battery is strictly confined to a group of several cells connected to form a convenient-sized unit of supply.

9.9.7 The Weston Standard Cell

This rather complicated cell has the advantage of maintaining a very steady and predictable e.m.f. (1.0186 V at 10°C). Its chief use is as a laboratory standard of e.m.f.; it cannot be used as a source of current (figure 9.51).

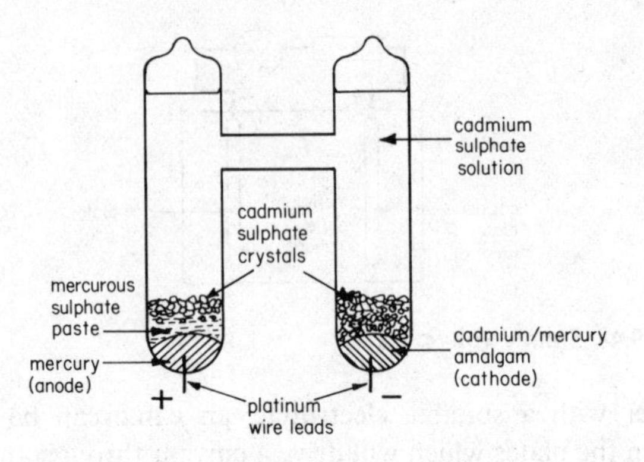

Figure 9.51 Weston cadmium cell

9.9.8 Primary and Secondary Cells

All the foregoing types of cell cannot be recharged when exhausted, except by the replacement of the consumed materials. These are classified as *primary cells.* Other types of cell exist which can be recharged by driving a reverse current through them, which restores the chemicals of the cell to their initial condition ready for a new working cycle. These are called *secondary cells* or accumulators.

The Lead–Acid Accumulator

The plates of this cell are frameworks made of lead which form the supports for *active material.* The active material for a *fully charged* cell is lead dioxide, PbO_2 (a dark brown paste), for the positive plate, and spongy lead for the negative plate (figure 9.52).

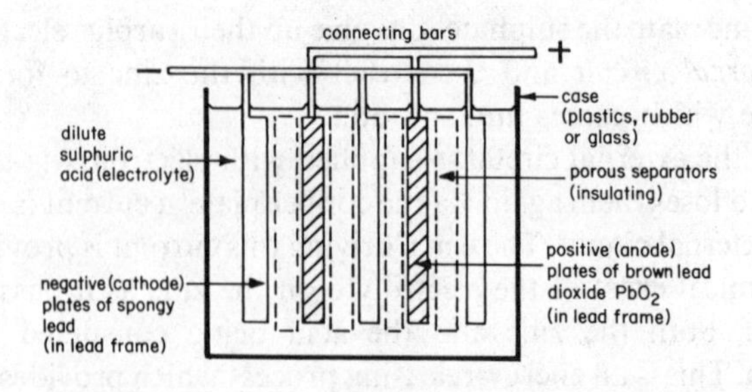

Figure 9.52 Lead–acid cell (cover and vents not shown)

The electrolyte is dilute sulphuric acid. During discharge *both* plates undergo a change to lead sulphate, which is white in colour. This process results in an increase in volume which, if allowed to occur too rapidly, will distort the plates. For this reason short circuits can cause serious damage to lead–acid cells and should be avoided. To prevent adjacent plates coming into contact, a *separator* made of porous insulating material is sandwiched between them.

The plates are not filled initially with the active material, but are first packed with cheaper red lead (Pb_3O_4) and litharge (PbO) for the positive and negative plates respectively. A *forming current* is then passed through the cell, which converts these materials to the required compounds.

The recommended maximum rates of charge and discharge must be carefully observed to avoid damage and a typical value is the 10-*hour rate*. This is the steady current which would completely discharge the cell in a time of 10 hours. The capacity of a lead–acid cell or battery is usually quoted in ampere-hours (1 A h = 3600 coulombs), and the 10-hour rate is numerically equal to one-tenth of the ampere-hour capacity, thus a 40-A h battery would be charged at a 10-hour rate of 4 A. Some cells and batteries are specially constructed to handle heavier currents but reference should always be made to the maker's specification before use. Lead–acid cells give reliable service with a reasonably long life, but must be treated with due respect.

During discharge the terminal voltage remains steady at 2 V for about two-thirds of the discharge time (assuming a 10-hour rate) but falls steadily after this to a value of 1.8 V when it quickly drops to zero (figure 9.53). Heavier discharge rates produce a more rapid decline with a less stable terminal voltage. The state of charge is difficult to interpret from the terminal voltage but is revealed by inspection of the electrolyte's specific gravity, which declines fairly steadily from 1.25 when fully charged to 1.18 when discharged.

The internal resistance of lead–acid cells is very low (see also section 9.3) which has advantages in some experimental work but results in extremely heavy currents when short circuited. Cells deteriorate by the spontaneous formation of lead sulphate if they are left discharged or unused for long periods. They benefit from regular charge/discharge cycles which should be arranged arti-

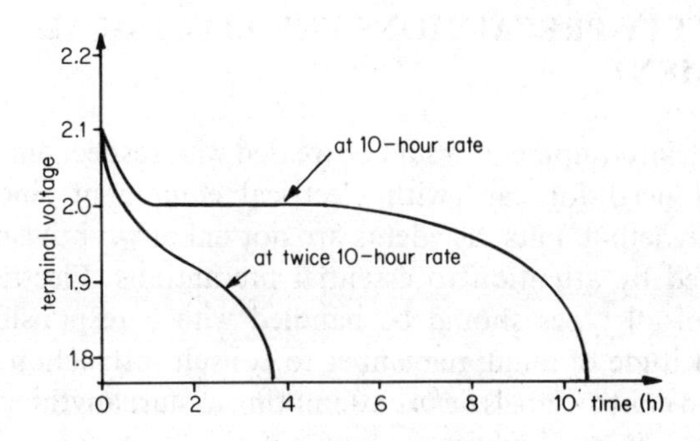

Figure 9.53 Discharge curves for lead–acid cell

ficially if necessary, or by using cells and batteries in orderly rotation.

The electrolyte is not consumed by the cell action but may lose part of its *water* content by evaporation. (The acid part will not evaporate easily.) The level should therefore be topped up with distilled water *only*. (*Care should always be exercised when handling acid, which becomes more corrosive as water evaporates after spillages.*)

Nickel–Iron (Alkaline) Cells

These use a potassium hydroxide electrolyte with active material composed of nickel oxide and hydroxide (anode) and iron oxide (cathode). These compounds are housed in robust perforated steel cases which are unharmed by heavy currents. They can be left idle indefinitely, are lighter in weight than the lead–acid type and of much more robust construction. On the other hand they have a lower e.m.f. per cell (1.2 V), cost more to make and have higher internal resistance with lower efficiency.

Both types of secondary cell have a very large mass per unit energy stored and intensive research is in progress to devise new methods of storage without this disadvantage, especially for the propulsion of vehicles.

9.10 SAFETY PRECAUTIONS FOR ELECTRICAL EQUIPMENT

All industrial equipment should be treated with respect but there is a special need for care with electrical equipment since it is potentially lethal. Fatal accidents are not unknown but can often be avoided by attention to essential precautions. Electrical apparatus of all types should be handled with a responsible and careful attitude of mind; remember to consult instruction panels, books and control labels *before* attempting to start anything—'If in doubt, ask!' is good advice.

Equipment supplied from the mains should be protected by fuses and circuit breakers. The ratings should be correct for the equipment and not increased to permit over-loading. When fuses and breakers have cut off the supply they should not be replaced until the cause of failure has been located and removed.

The condition of an unidentified conductor should be assumed dangerous until proved otherwise. Testing should be made with correct instruments and *not* with a finger!

All equipment should be kept in good condition by regular checks and inspections by suitably qualified persons.

The personal hazards encountered with electrical equipment can be grouped under the following headings.

9.10.1 Electric Shock

This is a form of *paralysis* which can affect heart and respiratory muscles, producing unconsciousness and death. It is caused by the conduction of quite small currents through the body (as low as 20 mA), from contact with an exposed conductor. The voltage required to produce this current is often over-estimated and conductors operating above 24 V should not be handled with unprotected hands.

The danger is increased if the skin is moist, if some other part of the body is well earthed, and if contact is maintained for an extended period. There is *no safe system* which can be handled with impunity.

Treatment demands *speed.* The current supply should be switched off, or the victim pulled away by an insulated material such as a *dry* piece of wood, cardboard or clothing. The victim's mouth should be cleared of obstructions such as dislodged false teeth, and his clothing loosened around the neck. Then artificial respiration should be given. All work places have a diagram showing the technique to be followed, which should be studied and practised as a routine by everyone likely to be involved. (It is too late to wait for the accident before consulting the diagram.) The treatment must be continued until the victim recovers or a *doctor* certifies death has occurred.

Apart from direct contact with an unprotected conductor, the following faults can also be the cause of electric shock.

(1) Inadequate earthing of equipment. Earthing requires a *continuous* path to a proper earth connection, and it may become interrupted by damage to the earth lead, even though the appliance functions normally. The continuity may be broken some distance away from the site of operations, where its significance may be overlooked.

(2) Use of flexible leads which are damaged by impact of tools or equipment. Leads should be of adequate quality and strength to withstand the rigours of normal use.

(3) Switchgear and implements used with covers removed or damaged. These should be repaired before use.

(4) Improvised connections to socket outlets and improvised cable joints are sometimes used by foolhardy people, who should be discouraged. Such assemblies should be carefully disconnected and replaced by correctly fitted items. Special connectors are available for temporary connections when testing equipment.

9.10.2 Burns

Electrical equipment can produce severe burns in the event of short circuits and electric arcs. Accidental arcs produce molten metal which can cause serious damage to skin tissue. Burns can be avoided by safe working methods and situations likely to involve short circuits and arcs should be anticipated; for example, 'testing' large accumulators by flicking a wire across the terminals is dangerous and harmful to the accumulator.

Burns should be treated by covering with a dry sterile dressing *only*, further help being obtained from a doctor. Fires started by electrical accidents should only be tackled by dry powder and/or CO_2-type extinguishers. Water-based appliances should not be used; read the label on the container.

9.10.3 Chemical Burns

Tissue damage is caused by contact with electrolytes of cells which may be acid *or* alkaline. Protective clothing including eye protection should *always* be used when handling these liquids and containers should be *correctly labelled.*

Chemical burns should be treated by flushing with cold water and applying a weak antidote. (Dilute vinegar for alkali burns, sodium bicarbonate solution for acid burns.) A doctor should be consulted as soon as possible.

9.10.4 Injuries caused by Electrical Machinery

Severe injuries can be caused by clothing becoming entangled in rotating machines, or by loose articles dropped into machines being ejected at high speeds (for example, pens and rulers falling from the pockets of overalls). These accidents are more likely to occur if machines are operated without protective guards. If guards and covers are removed for adjustments or maintenance they *must be replaced* before starting up.

Other precautions include care with magnets and electromagnets which exert large forces on steel objects in close proximity to the poles.

Accidental movements of tools or other steel objects can result in trapped fingers.

1 Charles A. Coulomb (1736–1806) was a French physicist. He invented a sensitive torsion balance which allowed him to measure electrostatic forces, leading to the identification of electric charge.
2 Andre M. Ampere (1775–1836) was a French mathematician and physicist. He studied the effects of current flow in conductors, particularly the electromagnetic effects. He devised the 'corkscrew' or right-hand screw rule for electromagnetic effects. He identified the force between two parallel conductors which is now used to define the ampere unit.
3 Alessandro Volta (1745–1827) was an Italian physicist. He devised experimental methods for studying electrostatics and devised the voltaic cell to produce currents of electricity, later used by Sir Humphry Davy and others.
4 Georg Ohm (1787–1854) was a German physicist. He investigated the relationship between potential difference and the flow of current; he defined resistance as the relating parameter in 1827.
5 Ernst Werner von Siemens (1816–92) was a German electrical engineer. He invented an electroplating process and the self-exciting dynamo.
6 James Watt (1736–1819) was a Scottish engineer. He made fundamental contributions to steam-engine design, resulting in its use as the power plant of the Industrial Revolution. He devised the governor for automatic engine control and introduced the horsepower as the unit of power measurement. (See also p. 92.)
7 Michael Faraday (see p. 167).
8 Georges Leclanche (1839–82) was a French chemist.

TO THE STUDENT

At the end of this chapter you should be able to

(1) understand the concept of electric charge
(2) recognise the behaviour of currents of charge moving through conductors
(3) understand the electric circuit and its property of resistance
(4) understand the factors which determine the resistance of a conductor
(5) be able to calculate the power and running cost of an electrical appliance
(6) understand the principles of simple electrical instruments in common use
(7) recognise the special treatment needed for measurement of alternating-current electricity

(8) understand the principles of electrostatics as applied to capacitance
(9) understand the basic principles of electrolysis and recognise the similarities between this process and cells, both primary and secondary
(10) appreciate the need for care and responsibility in the use of electrical apparatus
(11) complete exercises 9.1 to 9.15.

EXERCISES

9.1 A circuit is supplied with an electric current of 15 A for a period of 2 min. Throughout this time the p.d. across the ends of the circuit is 25 V. Calculate for the circuit (a) the charge, in coulombs, passing during the whole period, (b) the energy, in joules, consumed.

9.2 An e.m.f. of 2 V is used to drive a current of 0.5 A through an unknown resistance. Calculate (a) the resistance in ohms, (b) the current flowing if the e.m.f. is raised to 7.5 V, (c) the e.m.f. required to drive a current of 12 A through the same resistance.

9.3 Calculate the combined resistances of the groups of resistors shown in figure 9.54, and the current flowing in the circuit of figure 9.54c.

9.4 An electrical distribution system consists of four lengths of cable connected in parallel. If the conductance values of the cables are 20, 16, 40 and 32 S respectively, calculate the total conductance of the combined system and the total current flowing if the potential drop across the system is 1.2 V.

9.5 A cell of internal resistance 0.5 Ω is tested by a high-resistance voltmeter and is found to have a terminal voltage of 1.52 V. Calculate the voltage across the terminals when the cell delivers a current to an external load of 3 Ω.

9.6 (a) Calculate the resistivity of aluminium if the resistance of a conductor 10 m long with cross-sectional area 2 mm^2 is 0.132 Ω.

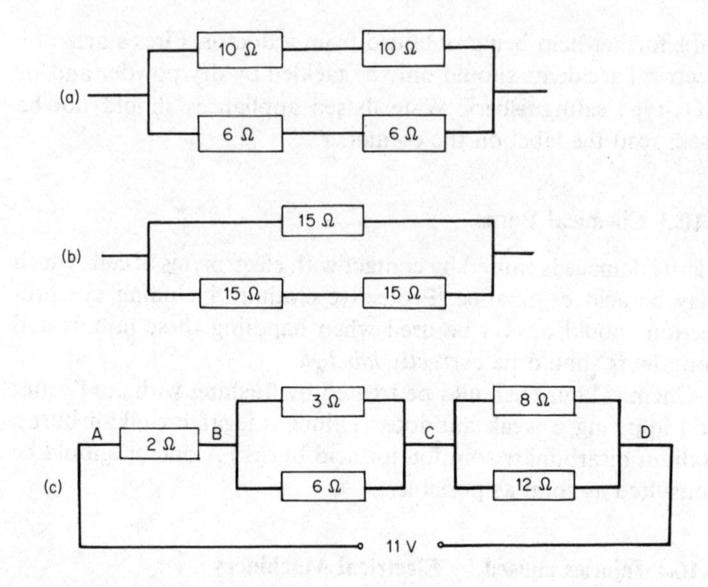

Figure 9.54

(b) What is the resistance of a conductor with a cross-sectional area 1.5 times that in (a) and length twice that in (a), made of the same material?

9.7 A metal resistance element has a measured value of 420 Ω at its working temperature of 120°C. The resistance was measured a second time at room temperature (20°C) and found to be 418 Ω. Calculate (a) the temperature coefficient of resistance for the metal, (b) the resistance value at 0°C.

9.8 If the cost of electricity is 1.4 p per unit, calculate the weekly cost of the following appliances used as indicated: electric iron (750 W) used on 3 days for 30 min each day, television set (250 W) used every day for 2 h, electric fire (2 kW) used once for 1.5 h, 5 lamps used every day for 5 h.

9.9 Calculate the power rating of the following devices: (a) an electric heater connected to a 240-V supply and using a current of

40 A, (b) a resistor of value 47000 Ω carrying a current of 2 mA, (c) a heater of resistance 60 Ω connected to a supply of 240 V.

9.10 A moving-coil meter has an internal resistance of 35 Ω and requires 2 mA for full-scale deflection. Calculate suitable shunts to enable it to measure (a) 20 A f.s.d., (b) 10 A f.s.d. The same meter must also be able to measure 400 V f.s.d. Calculate a suitable multiplier resistance and show how it should be connected in circuit with the meter movement.

9.11 A resistance is to be measured by the ammeter/voltmeter method. The ammeter resistance is 0.1 Ω and the voltmeter resistance is 2500 Ω. Draw a circuit arrangement and calculate the error involved if the true value of the resistance is 2000 Ω.

9.12 An alternating-current supply has a sinusoidal waveform with a peak voltage value of 350 V. A resistor of 100 Ω is connected across the supply for use as a heater. Calculate (a) the mean voltage, (b) the r.m.s. current, (c) the power rating.

9.13 A capacitor is formed by two aluminium foil sheets 20 cm square separated by a polythene sheet 0.07 mm thick. Calculate the value of its capacitance if the permittivity of free space is 8.85×10^{-12} F/m and the relative permittivity of polythene is 2.3.

If a potential of 40 V is maintained across the plates, calculate the charge stored by the capacitor and the energy required to maintain the charge.

9.14 A copper-plating cell deposits a mass of 2 g of copper during a period of 2 h. If the electrochemical equivalent of copper is 0.329×10^{-6} kg/C, calculate the steady current flowing. How long would the same current have to flow to deposit a mass of 5 g of silver (e.c.e. for silver = 1.118×10^{-6} kg/C)?

9.15 A lead–acid accumulator has a capacity of 25 A h. (a) Calculate the charging current required at the 10-h rate. (b) What is a reasonable maximum discharge rate if the design maximum is not known? (c) The specific gravity on test is found to be 1.18. What does this reveal about the state of charge? (d) What action should be taken if the level of the electrolyte has diminished by 10 mm below the indicated level?

NUMERICAL SOLUTIONS

9.1 (a) 1800 C, (b) 45 kJ
9.2 (a) 4 Ω, (b) 1.875 A, (c) 48 V
9.3 (a) 7.5 Ω, (b) 10 Ω, (c) 8.8 Ω, 1.25 A
9.4 108 S; 129.6 A
9.5 1.283 V
9.6 (a) 2.64×10^{-8}, (b) 0.176 Ω
9.7 (a) 45×10^{-6}/°C, (b) 417.74 Ω
9.8 35.2 p
9.9 (a) 960 W, (b) 0.188 W, (c) 960 W
9.10 (a) 0.0035 Ω, (b) 0.007 Ω; multiplier value 199 965 Ω (series)
9.11 Error = 0.1 Ω = 0.005%
9.12 (a) 223 V, (b) 2.47 A, (c) 610 W
9.13 0.012 μF; 0.465×10^{-6} C
9.14 0.84 A; 1 h 28 min 19 s
9.15 (a) 2.5 A charge, (b) 2.5 A max discharge

10 Magnetism and Electromagnetism

The object of this chapter is to give the student an understanding of the effects relating magnetism and electricity and to observe the various ways this is applied in engineering.

10.1 EVIDENCE OF A MAGNETIC EFFECT

The effect of mutual influence without the need for an intervening medium, known as *gravity*, was recognised early in history and clearly understood from the time of Newton. However, the apparently similar effect produced by magnetic materials was less frequently encountered and consequently less well understood until more recent times. The effect was first observed in certain natural materials containing a special type of iron ore; small pieces were mutually attracted or repelled, undoubtedly under the action of a *force*.

Among the special properties of these natural magnets was the ability to return to a north–south disposition when freely suspended, a property quickly adopted for navigation. These pieces of ore were known as 'lodestones' and enabled journeys to be made by more reliable guidance methods than hitherto. This apparently strange behaviour is now recognised as being caused by the magnetic influence of the Earth itself, which is in effect a large magnet, possibly due to large amounts of magnetic material in its structure.

Techniques are now available to make magnets artificially. It is now readily acknowledged that a magnetic effect exists, consisting of forces (often attractive but sometimes repulsive) exerted on certain kinds of object.

10.1.1 Magnetic Materials

The natural material of the early magnets was black iron oxide which has magnetic properties. The purified oxide is still used for magnets, for example, the ring magnets used for focusing the electron beam in television cathode-ray tubes.

It was found quite quickly that the natural magnets could make iron and steel objects become magnetic and in some cases retain their magnetism. Steel is a much more durable material than crumbly iron oxide and so was used for making the compass 'needles' used in place of lodestones.

The usual modern magnetic materials are the elements iron, cobalt and nickel in the metallic state, or alloys containing these metals. Because of their association with iron and its properties, these are called *ferromagnetic* materials. The black iron oxide (ferrite) mentioned earlier is also used, bonded with synthetic resins or ceramics to give better mechanical strength. This material is a *ferrimagnetic* material and its fundamental behaviour differs from that of ferromagnetic material but outwardly it is the same.

We should note here that the material alone is not necessarily a magnet, but it may acquire magnetic properties when subject to appropriate treatment. Some materials may retain their magnetic effects for long periods and are called *permanent magnets*; usually these are *hard* metals, steels and similar alloys of iron. Other materials lose their magnetism soon after the magnetising treatment is removed; soft iron is an example, as are special alloys developed for applications where magnetism is frequently created then destroyed, such as transformer cores (see section 10.7).

Typical Magnetic Materials

Material	Type	Use
Iron Nickel Cobalt Stalloy (iron– silicon) Mumetal (complex alloy)	Soft	Cores and armatures for transformers and electrical machinery
Carbon steel Cobalt steel Ticonal alloy Alnico (Alcomax) alloy	Hard	Permanent magnets used in meters, microphones, loudspeakers, etc.

Certain other materials display minor magnetic properties but with the exception of those mentioned, most materials are non-magnetic.

Magnets can be made in a variety of shapes, depending on the applications, but one of the commonest is the bar magnet. If such a bar is bent into a U shape it is called a 'horseshoe' magnet. Others may be in the form of rings or more complex configurations.

10.1.2 Behaviour of the Magnetic Force

It has been observed that magnets exert their force on magnetic material some distance beyond their own boundary surfaces and the term *magnetic field* describes this region of influence (figure 10.1). The field is strongest close to the magnet but rapidly weakens with distance until eventually the effect is negligible, depending on the strength of the magnet (figure 10.2).

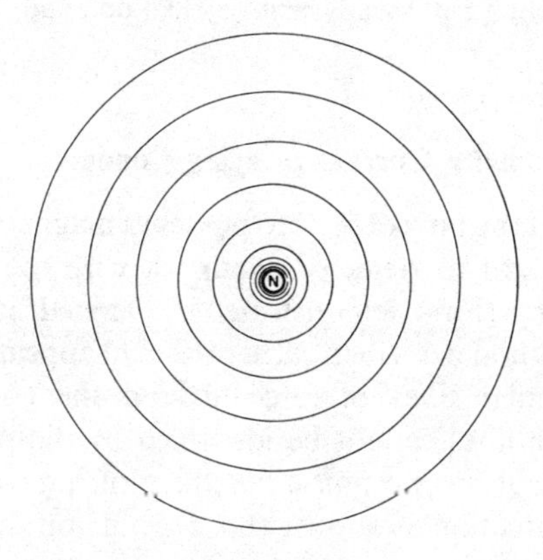

Figure 10.1

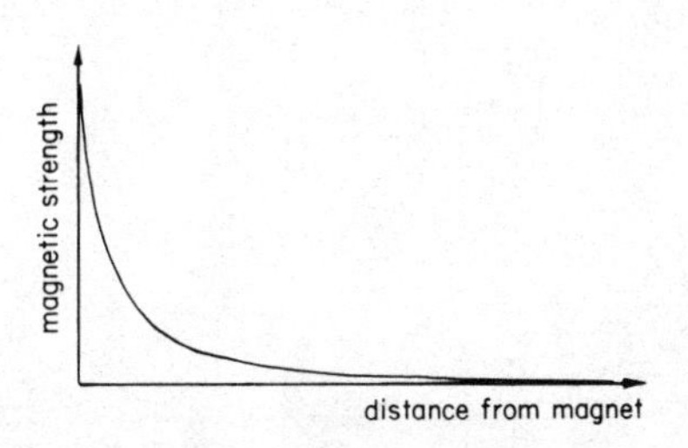

Figure 10.2 Variation of magnetic strength

Forces are detectable between two magnets as their fields interact, and the direction and type of force depend on the *polarity* of the magnets. It has been noted earlier that a suspended magnet points in a north–south direction and we may describe the extremities as north-seeking and south-seeking 'poles', or *north pole* and *south pole* for short. If two magnets are placed with their north poles near each other the magnetic force produced is *repulsive*, that is, the magnets try to push each other away; but if the north pole of one is placed near the south pole of another the force is *attractive*, trying to draw the magnets together (figure 10.3). These tests show that magnetic fields possess a *directional quality*.

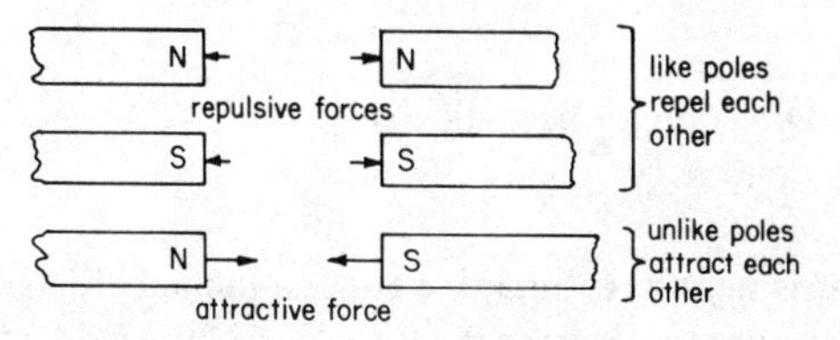

Figure 10.3

10.2 SHAPE AND DISTRIBUTION OF MAGNETIC FIELDS

Experiments can be conducted to 'plot' the direction of a magnetic field (figure 10.4) using small 'compass' needles (small magnets suspended on a pivot). The resulting field for a bar magnet is as shown in figure 10.5. There appears to be a series of lines travelling from one pole to another in continuous loops but in fact these are only the paths *actually plotted*. The field is *continuous between the plotted paths*; the lines only indicate the direction of the field.

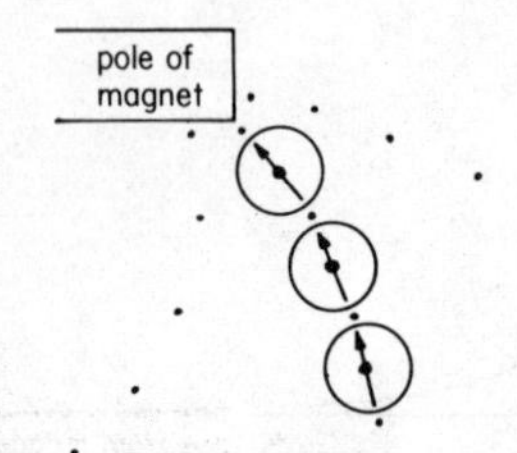

Figure 10.4 Plotting technique using compass

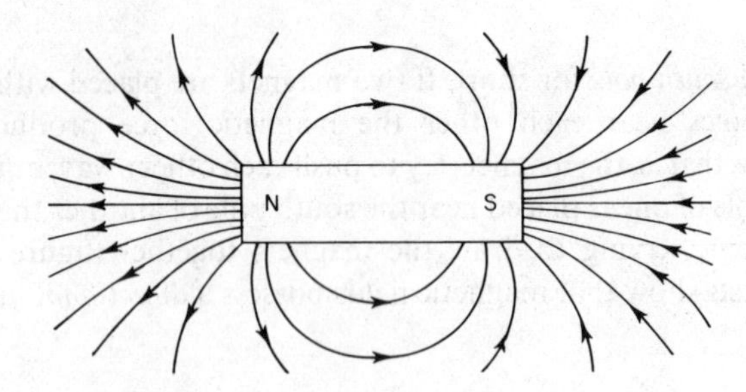

Figure 10.5 Direction of magnetic field

However, it is helpful to imagine the magnetic field as made up of 'lines'. We choose to describe the direction of the field as being from a north pole to a south pole, as shown by the arrows in the diagram.

10.2.1 Combined Fields

The fields around two magnetic poles brought close together (figure 10.6) combine in such a way as to support the theory that attraction and repulsion forces depend on the magnetic field. We may imagine the 'lines' of magnetism (also called lines of force) behave as if they were stretched fibres which in some way repel their neighbours and are in a state of tension, trying to become as short as possible (a concept put forward by Faraday).[1] The lines readily travel from a north pole to a south pole but not from north to north or south to south.

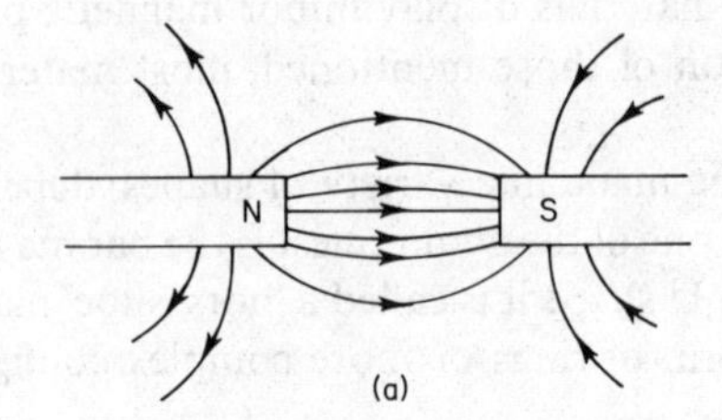

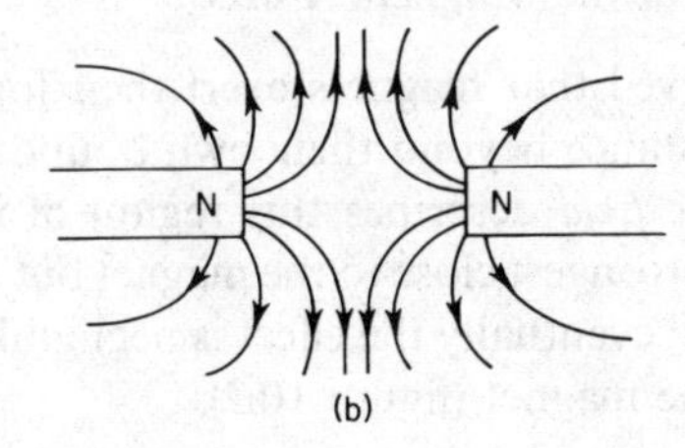

Figure 10.6 (a) Unlike poles adjacent, (b) like poles adjacent

10.2.2 Field around a Current-carrying Conductor

Magnets are not the only objects to possess a magnetic field around them—a field can be detected around a wire through which a current is passing. (First demonstrated by Oersted[2] in 1820.) Using the plotting method described earlier the field appears to consist of concentric circular lines of force around the conductor. The direction of the lines cannot be identified by the polarity of the conductor since it has no poles, but the plotting compass needle indicates the direction as shown and depends on the direction in which the current is flowing. If the current is reversed, the direction of the lines is reversed. A useful rule is the 'corkscrew' rule which states that the direction of the current is represented by the *penetration* of a corkscrew and the direction of the field lines by the associated rotation (figure 10.7).

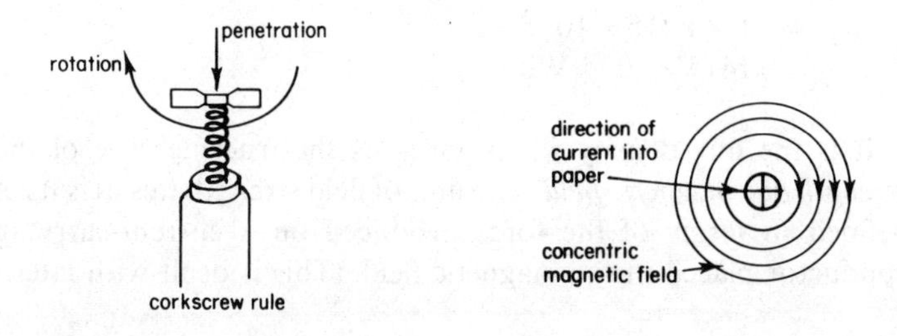

Figure 10.7 Field due to a current

10.3 THE CONCEPT OF MAGNETIC FLUX AND FIELD STRENGTH

There are two main factors which affect the strength of the magnetic effect.

(1) The strength of the source of the magnetism, conductor or magnet, which can have various magnitudes.
(2) The distance from the source at which the effect is measured. This is because the lines spread out as we move away from the source.

We might expect therefore that the magnetic effect in a field is not necessarily a constant quantity. The magnetic effect of the field depends on the field configuration and direction as indicated by the lines of force which are distributed in patterns depending on the local circumstances (see section 10.2). Sometimes these are close together and at other times far apart and we have to take account of the amount of magnetic effect passing through some observation 'window' at a given location (figure 10.8). We call this quantity the magnetic flux (symbol ϕ). (Note that, although flux means flow, there is no physical flow involved here.) The unit for measuring flux is called the *weber* (Wb).[3] The precise definition of the weber is given later.

Another useful quantity in this section is the *flux density* (symbol *B*) which tells us how the flux is distributed. If the flux is concentrated in a small area we say it has a high flux density, but the *same flux* spread out over a large area has a low flux density. To measure flux density we divide the flux in webers by the area in

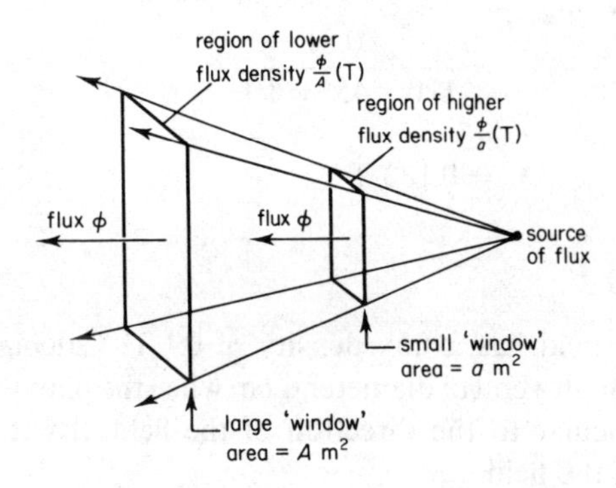

Figure 10.8 Variation of flux density

metres^2 *measured perpendicular to the direction* of the flux (figure 10.9). The unit is called the tesla (T).[4] Thus

$$\text{flux density } B = \frac{\text{flux } \phi \text{ (Wb)}}{\text{area } A \text{ (m}^2\text{)}} \text{ T}$$

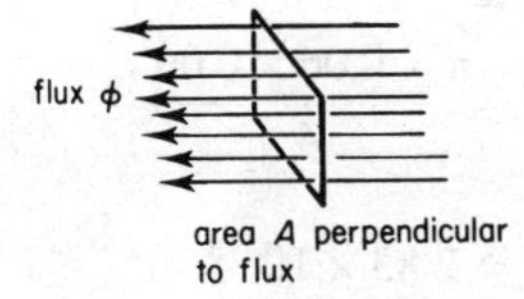

Figure 10.9

Example 10.1

Calculate the flux density produced by a flux of 0.01 Wb concentrated in an air gap of cross-section dimensions 120 mm by 455 mm.

Solution

$$\text{Flux density, } B = \frac{\text{flux } \phi}{\text{c.s.a } A}$$

$$= \frac{0.01}{120 \times 455 \times 10^{-6}}$$

$$= 0.183 \text{ T}$$

Example 10.2

A magnetic field has a flux density of 0.1 T. Calculate the flux passing through a coil of diameter 6 cm when the plane of the coil is (a) perpendicular to the direction of the field, (b) at 30° to the direction of the field.

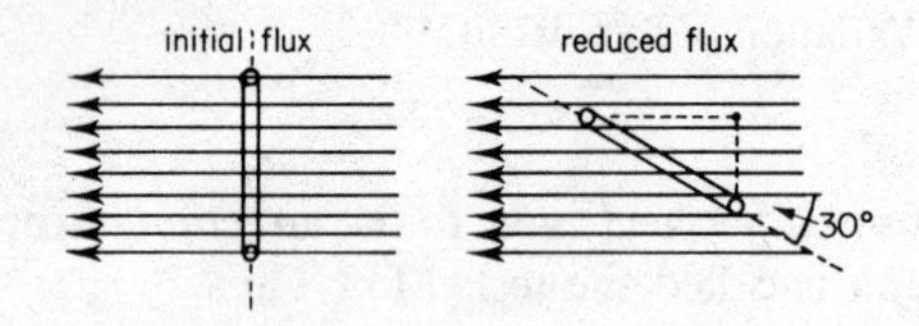

Figure 10.10

Solution (see figure 10.10)

$$\text{Area of coil } \frac{\pi d^2}{4} = \frac{\pi \times 0.06 \times 0.06}{4}$$

$$= 2.83 \times 10^{-3} \text{ m}^2$$

(a) When coil is perpendicular to field

$$\begin{aligned} \text{flux } \phi &= \text{flux density } B \times \text{c.s.a } A \\ &= 0.1 \times 2.83 \times 10^{-3} \\ &= 0.283 \times 10^{-3} \text{ Wb} \end{aligned}$$

(b) When coil is at 30° to field the c.s.a. penetrated by the flux is

$$\begin{aligned} A &= 2.83 \times 10^{-3} \sin 30° \\ &= 1.415 \times 10^{-3} \text{ m}^2 \end{aligned}$$

hence

$$\begin{aligned} \phi &= 0.1 \times 1.415 \times 10^{-3} \\ &= 0.1415 \times 10^{-3} \text{ Wb} \end{aligned}$$

It is the flux density which gives us the true measure of the *strength of a magnetic field*. The unit of field strength (tesla) is itself defined in terms of the force produced on a current-carrying conductor placed in the magnetic field. (This is dealt with later.)

10.3.1 Intensification of Flux Density

Fields of high flux density are often required, for example, in lifting magnets and electrical machines, and this can be achieved by two methods, which are often used together.

The Solenoid or Coil

The field around a conductor is in the form of concentric circles but if the conductor is wound into a circle the flux within the circle is increased (figure 10.11). By using many such 'turns' we can obtain a very large flux in the same area, that is, a high flux density. By

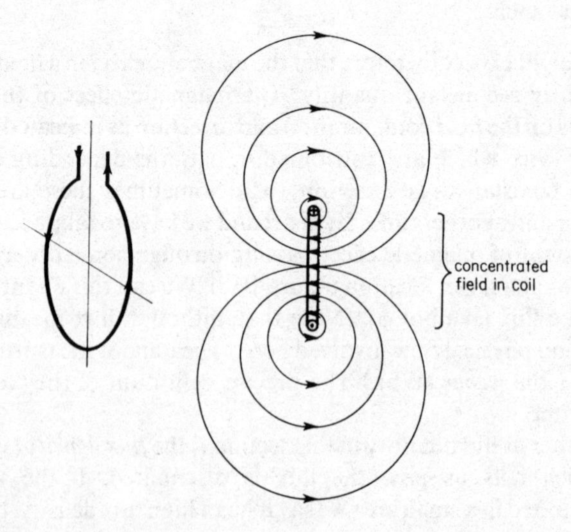

Figure 10.11 Magnetic field of a coil

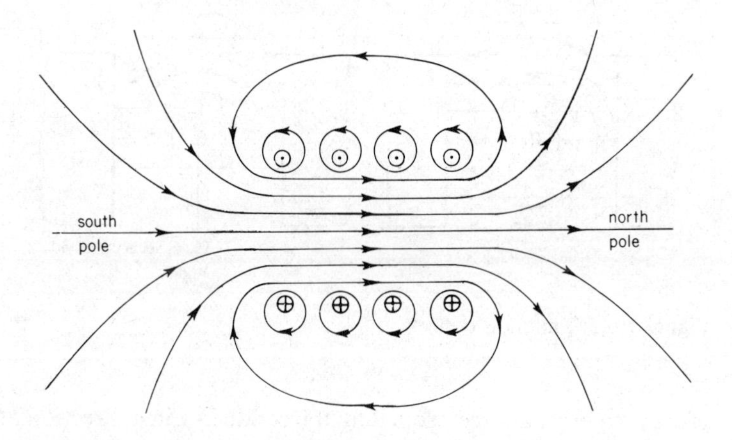

Figure 10.12 Solenoid field

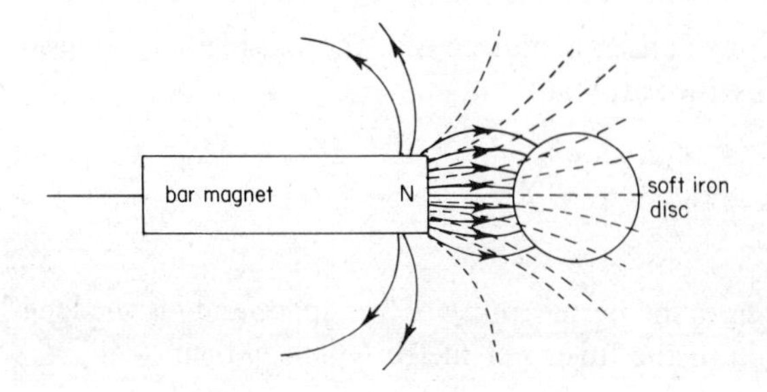

Figure 10.14 Effect of iron

extending the coil so formed into a cylinder we can produce a field parallel to the axis of the coil, which will resemble that of a bar magnet, complete with north and south poles (figure 10.12). The polarity can be determined by inspection of the diagram, but a useful rule is to look into the end of the coil and note the direction of current flow. Clockwise indicates a south pole and anticlockwise a north pole. (The rule is remembered easily by writing N and S with arrow heads on the ends of the letters—figure 10.13.) A magnet formed in this way is called an *electromagnet* and the actual coil is called a solenoid.

Figure 10.13 Solenoid polarity

Use of an Iron Core or Magnetic Path

If the field is observed around a magnet it has a familiar shape but this becomes distorted if a piece of iron is introduced (figure 10.14). It appears that the lines of force which define the flux distribution prefer to align themselves *through the iron* rather than the surrounding air, even increasing in length to do so in some cases. To explain this behaviour we suppose that it is easier for the flux to pass through iron than through the air and the flux adopts the path of least resistance. The ease with which flux penetrates a material is called the *permeability* of the material (symbol μ). Iron has a high permeability, air has a very low permeability, and free space is even lower (the symbol for the permeability of free space is μ_0). Inspection of figure 10.14 shows that the piece of iron has a much larger flux density than the surrounding area and tends to concentrate the flux. This effect is widely used in electromagnets for motors, generators and other equipment where a concentrated flux is required.

10.3.2 Field Strength Produced by an Electromagnet

A solenoid produces a flux inside its coils which is related to the following values: (1) the current in the conductor, I, (2) the number of turns on the coil, N. The product of these two, NI, is called *ampere-turns* and determines the magnitude of the agency which produces the flux and 'drives' it through the media which form its path. For this reason the ampere-turns quantity is often referred to as the *magnetomotive force*, or m.m.f., analogous to the electromotive force in current electricity. (Note that neither is a true force in the mechanical sense.)

The flux density produced near the axis of a very long solenoid is given by the formula

$$B = \frac{\mu_0 N I}{l} \text{ T}$$

where μ_0 is the permeability of free space and l is the length of the solenoid. If the turns per metre $N/l = n$, then

$$B = \mu_0 n I \text{ T}$$

($\mu_0 = 4 \times 10^{-7}$ henrys per metre; the henry is discussed in section 10.8).

Example 10.3

Calculate the flux density of the field in the centre of a long coil with 500 turns if the dimensions are: length = 45 cm, diameter = 5 cm, and a current of 10 A is flowing in the coil (assume $\mu_0 = 4 \times 10^{-7}$ H m^{-1}).
Solution

$$\text{Flux density} = \frac{\mu_0 N I}{l} \text{ T}$$

(note that diameter has no effect)

$$= \frac{4 \times 10^{-7} \times 500 \times 10}{0.45}$$

$$= 0.044 \text{ T}$$

10.3.3 Combined Field for a Conductor placed in a Magnetic Field

If we superimpose the two fields we find an area of weak field on one side of the conductor where the lines are in opposition, and strengthened field on the opposite side where the lines are combined. Remembering that the lines try to become shorter, the

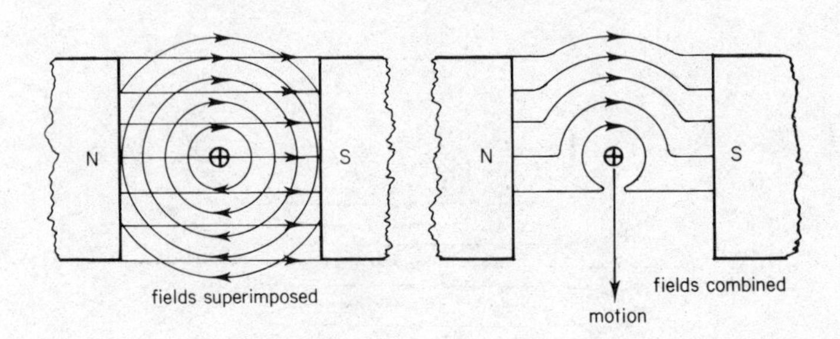

Figure 10.15 The motor effect

effect is to exert a force propelling the conductor in a direction at right-angles to the field, (often called the motor effect)—see figure 10.15. Inspection of the current direction and the directions of field and motion shows that they are mutually perpendicular. *Fleming's*[5] *left-hand rule* (figure 10.16) is a useful aid to remembering these correctly. If the thumb, forefinger and middle finger of the left hand are set mutually perpendicular they represent the motion, field and current directions respectively (thuMb = Motion, Forefinger = Field, mIddle finger = I (current).

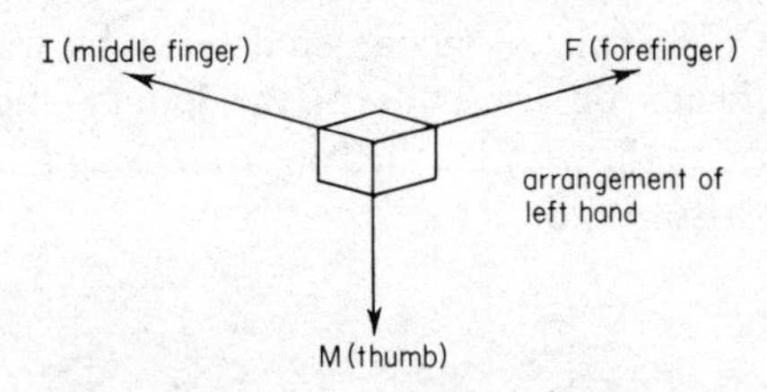

Figure 10.16 Fleming's left-hand rule

10.3.4 Magnitude of the Force exerted on a Current-carrying Conductor

The size of this force can be calculated from

(1) the flux density at right-angles to the conductor, B T

(2) the current in the conductor, I A
(3) the length of the conductor, l m.

The force is directly proportional to each of these terms, that is

$$F \propto B \qquad F \propto I \qquad F \propto l$$

It can be shown that the magnitude of the force is given by

$$F = BIl \text{ N}$$

Example 10.4

(a) A conductor 15 cm long lies between the poles of a magnet in a uniform field of 0.05 T. Calculate the force on the conductor when a current of 7.5 A is flowing. (b) If the arrangement is as shown in figure 10.17, determine the direction of motion produced by the force. (c) What happens to this direction if (i) the direction of field is reversed, (ii) both field and current are reversed.

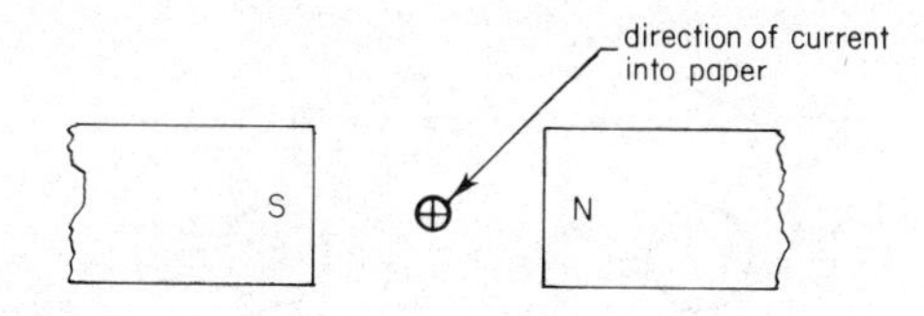

Figure 10.17

Solution

(a) $F = BIl$
$= 0.05 \times 7.5 \times 0.15$
$= 0.056$ N

(b) By application of Fleming's left-hand rule the force will be in an upward direction on the arrangement shown.

(c) Further use of Fleming's left-hand rule shows that (i) direction of force is reversed, (ii) direction of force is unchanged.

Definition of Flux Density Value

The value of B is defined in terms of the force produced on a long current-carrying conductor at right-angles to the field (figure 10.18). If the current is I A, the force F N and the length l m, then

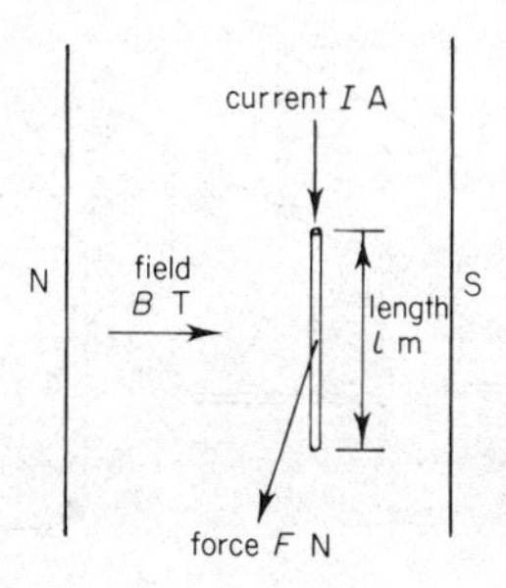

Figure 10.18

$$B = \frac{F}{Il} \text{ T}$$

(see also section 10.4.2, definition of the ampere). If we define field strength or flux density this way, we can then find the flux ϕ in webers through a particular area A, where

$$\phi = BA \text{ T m}^2 \text{ or Wb}$$

10.4 THE DIRECT-CURRENT MOTOR

A motor consists essentially of two conductors joined at the ends to form a rectangular loop or single-turn coil and positioned with two sides perpendicular to the direction of a magnetic field (figure 10.19). If a current is passed through the loop as shown, each conductor will be subject to a force according to Fleming's left-hand rule. These two forces are in opposite directions so that the loop is subject to a *torque*, and if pivoted along the axis of the loop it will rotate. This is the basis of the direct-current electric motor.

As the coil continues to rotate two problems emerge (figure 10.20).

(1) The perpendicular distance of the conductors from the

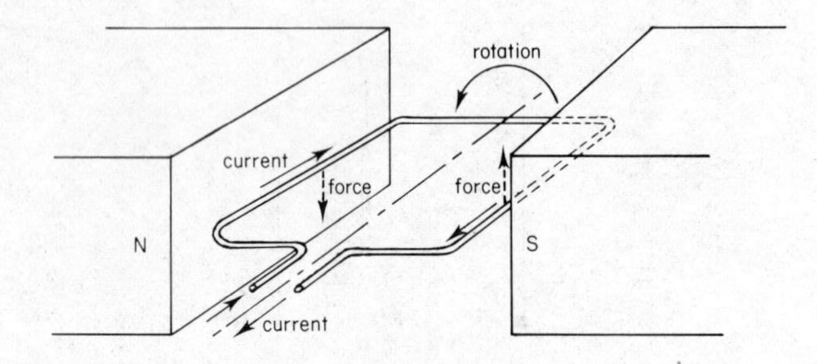

Figure 10.19 Basic design of electric motor

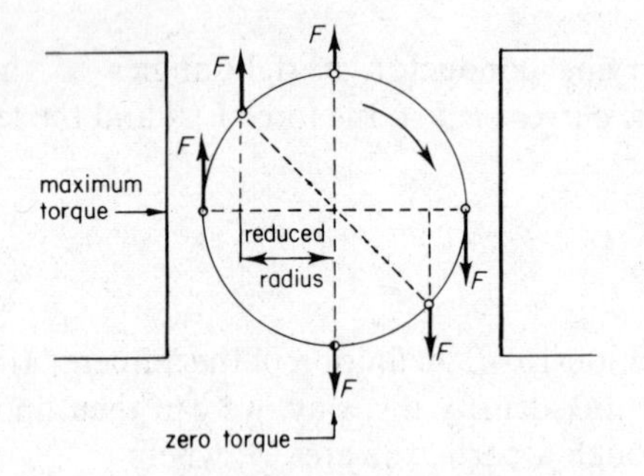

Figure 10.20 Variation of torque with angle of coil

centre of rotation diminishes, hence the torque is also reduced until, when the conductors are both in a plane at right-angles to the field, the torque is zero.

(2) After this point the motion of the conductors opposes the original direction of the torque, so the motor will stall or cease to rotate.

The first problem is overcome by the use of a *radial field* (figure 10.21) produced by placing a cylindrical iron armature in between shaped pole pieces (see section 10.3.1) and allowing the conductors to move in the gap between. Since the field is now radial the conductor always moves at right-angles to it and the torque remains constant.

In the case of the opposed direction of the torque after the 90° position, the obvious cure is to try to reverse the current so that the

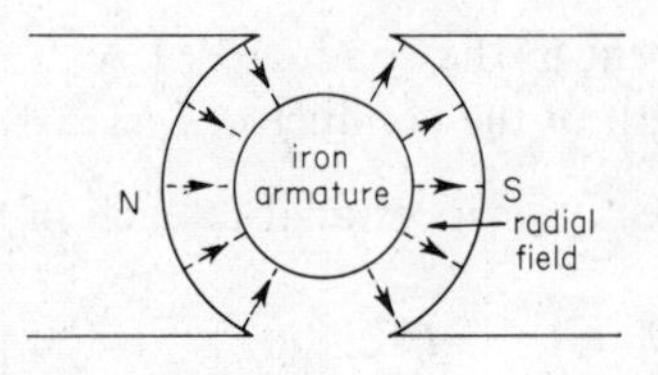

Figure 10.21 Creation of radial field

torque remains in the same direction for the whole revolution. This is achieved with an automatic changeover switch called a *commutator* (figure 10.22). Two sliding brushes, usually carbon, supply the current to separate contacts each forming half of a small cylinder attached to the motor shaft and rotating with it. As the diagram shows, the current is reversed every half revolution at the instant the conductor is about to re-enter the field. Hence the *direction* of the torque remains the same throughout the full revolution.

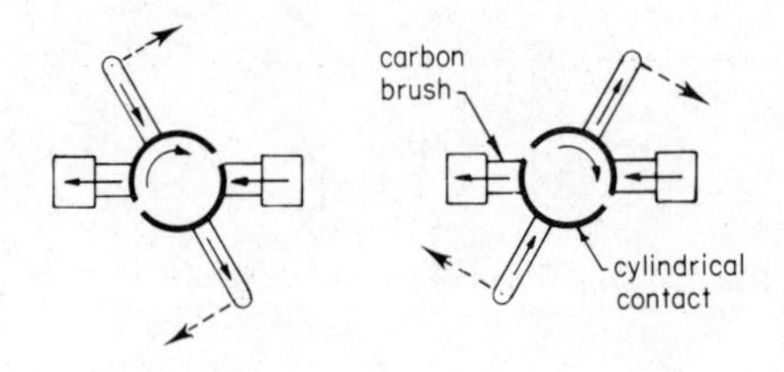

Figure 10.22 Use of commutator to reverse current

Example 10.5

A flat square coil of side 20 cm is wound with 50 turns of wire and is pivoted on a shaft as shown in figure 10.23. Calculate the torque exerted on the coil when a current of 5 A is flowing, the coil is located in a parallel magnetic field of 0.1 T, and (a) when the plane of the coil is parallel to the field, (b) when the plane of the coil makes an angle of 60° to the field, (c) when the plane of the coil has turned through 135° from its initial position.

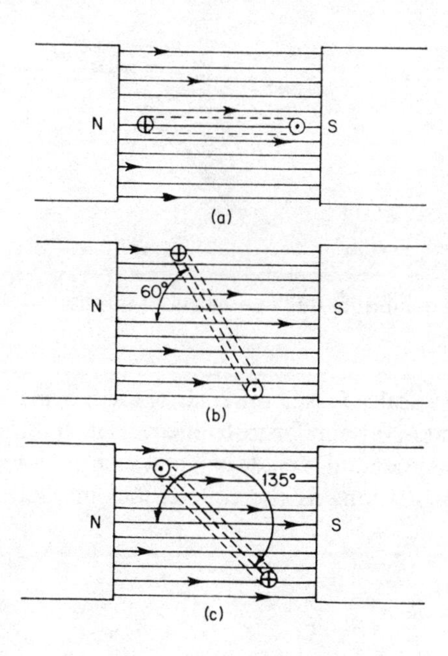

Figure 10.23

Solution By application of Fleming's left-hand rule we observe that opposite forces are exerted on the two sides of the coil at right-angles to the field. There is no force on the sides which lie parallel to the field.

Force on *one* side of coil $= BIl$
$= 0.1 \times 50 \times 5 \times 0.2$
$= 5$ N

(a)

Torque produced by couple with plane of coil parallel to field $=$ magnitude of one force $\times$ separating distance

$= 5 \times 0.2$

$= 1.0$ N m

(b)

Torque produced when coil is at 60° to field $= 5 \cos 60 \times 0.2$
$= 5 \times 0.5 \times 0.2$
$= 0.5$ N m

(c)

Torque produced when coil is at 135° to field $= 5 \cos 135 \times 0.2$
$= 5 \times (-0.707) \times 0.2$
$= -0.707$ N m

The direction of the torque has been *reversed.*

10.4.1 Applications of the Simple D.C. Motor Mechanism

The Motor as a Prime Mover

The simple arrangement described in the previous section is too weak to produce much power on its own, unless very large currents are used. To avoid the need for this, each loop is wound with many turns of wire (figure 10.24) so that quite modest currents can produce appreciable torques, since each turn of wire will make its contribution to the total torque. The torque can also be increased and smoother running obtained if there are several such coils arranged in different planes, each with its own pair of contacts on the commutator drum (figure 10.25).

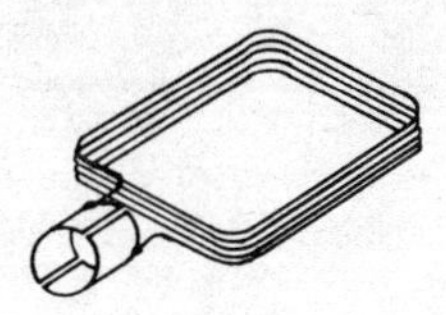

Figure 10.24 Multi-turn coil

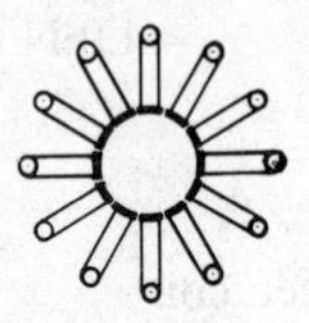

Figure 10.25 Multi-segment commutator

The Moving-coil Meter

The motor effect is used in many electrical measuring instruments. A coil of many turns is wound on a former and rotates on bearings in the radial field of a permanent magnet as described earlier (see figure 10.26). However, the rotating coil is attached to the end of a helical spring, the other end of which is fixed to the frame of the instrument and cannot move. When a current flows through the coil it experiences a torque which causes the coil to rotate, but as it does so the helical spring is deflected and eventually balances the motor torque (figure 10.27). The coil is unable to move further and reaches an equilibrium position which depends on the size of the current flowing. A heavier current will cause the spring to deflect a greater distance and a lower current a smaller distance. We can attach a pointer to the coil to amplify the movement, which is then

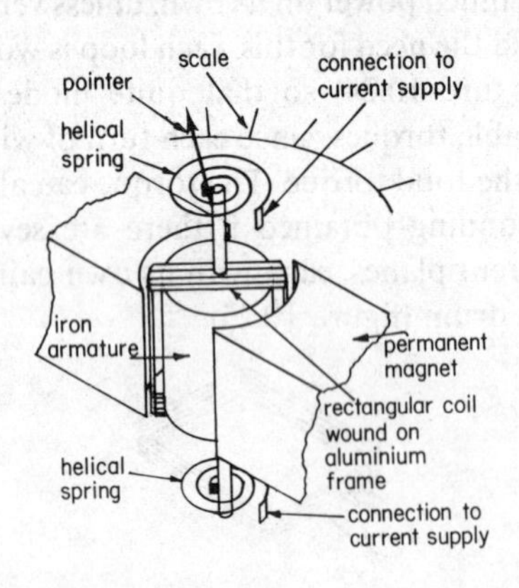

Figure 10.26 The moving-coil meter

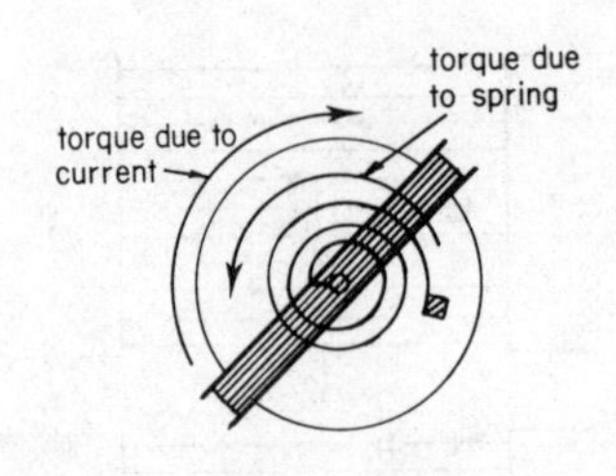

Fgure 10.27 Equilibrium due to a balanced torque

observed on a scale. This is the basis of the moving-coil ammeter, voltmeter and many similar instruments. The frame or former on which the coil is wound is made of metal such as aluminium to help damp out oscillations as the coil reaches its final position (see section 9.6).

The Loudspeaker

The moving-coil loudspeaker makes use of the motor effect to convert electric current variations into mechanical movements of a paper cone which set up audible pressure changes in the surrounding air (figure 10.28).

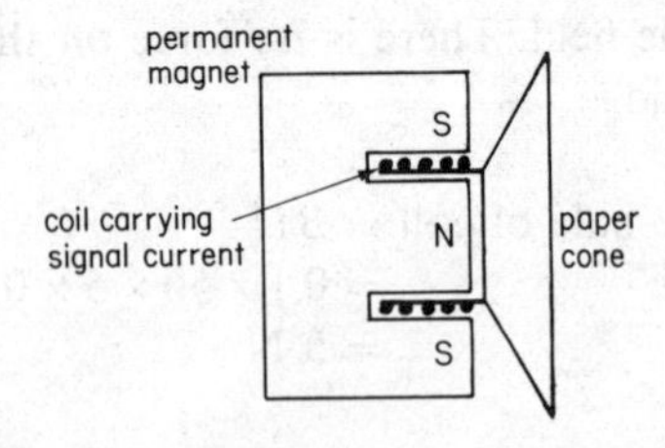

Figure 10.28 Moving-coil loudspeaker

The signal currents are fed through a cylindrical coil wound on a suitable former attached to the cone and located in the air gap of a permanent magnet with a very high flux density. An increase in the current causes an increase in the force on the coil acting in a direction given by Fleming's left-hand rule, that is, tending to eject the coil from the gap. Since the current is alternating in direction, the cone vibrates backwards and forwards in time with the variations in current and faithfully reproduces the original sound from which the currents were formed in the microphone.

10.4.2 Using the Motor Effect to Define the Ampere

The motor effect can be used to measure an electric current by observing the mechanical *force* produced. Thus it is not necessary to use any items of electrical apparatus and the measurement is reduced to an extremely simple form. We can set up the test as follows.

Current is passed through two very long parallel conductors located 1 m apart in a vacuum. Each conductor sets up its own concentric magnetic field according to the corkscrew rule.

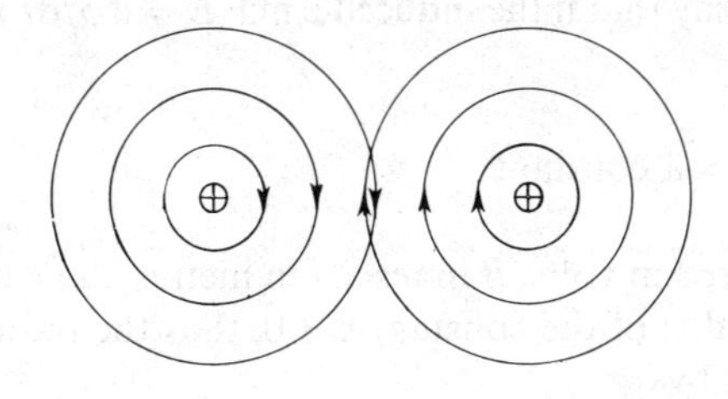

Figure 10.29

In each case the conductor now carries a current while lying in the field produced by the other conductor (figure 10.29), so each will experience a force according to Fleming's left-hand rule (figure 10.30). If both currents are in the *same* direction the forces will be *attractive*; if in *opposed* directions the forces are *repulsive*. The forces will always be in mutually opposite directions.

If the current is 1 A in each conductor the force exerted is *2* $\times$ *10^{-7} N/m length* of the conductors. This can be measured in special apparatus called a *current balance*, although in practice this device is somewhat different from the basic apparatus described above. (The force is very small.)

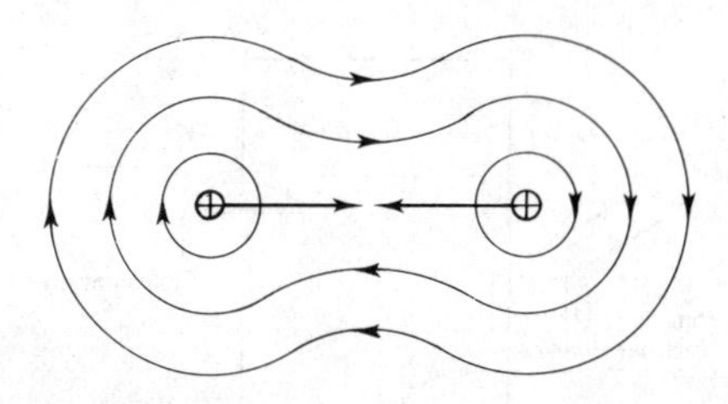

Figure 10.30

The value 2×10^{-7} N is an arbitrary value chosen so that the ampere defined in this way is similar in magnitude to the ampere defined by an earlier method, now disused; so there was no need to recalibrate electrical instruments already in use.

10.5 ELECTROMAGNETIC INDUCTION

One of the effects we have discussed in some detail is the force tending to move a current-carrying conductor lying in a magnetic field. As with many other effects encountered in science, this effect is reversible, and if a conductor is *moved* across a magnetic field, current can be made to flow in the conductor. This relative motion can be achieved by moving the conductor or the magnet. The current will only flow, of course, if the conductor forms part of a circuit but we can recognise the existence of an e.m.f. across the ends of the conductor even if there is no external circuit. The e.m.f. is said to be *induced* in the conductor by its movement across the field, and the effect is known as *induction*.

A simple demonstration is to move a bar magnet past a straight wire connected in circuit with a sensitive galvanometer (figure 10.31). As the magnet moves past the wire, the needle gives a flick but the effect is too small and short-lived to measure easily. The effect is more marked if we use a coil instead of a wire and insert the magnet into the coil. *While the magnet is moving* the meter shows a more noticeable swing. A significant feature of this test is that the swing of the meter is greater if the *speed of insertion* is increased.

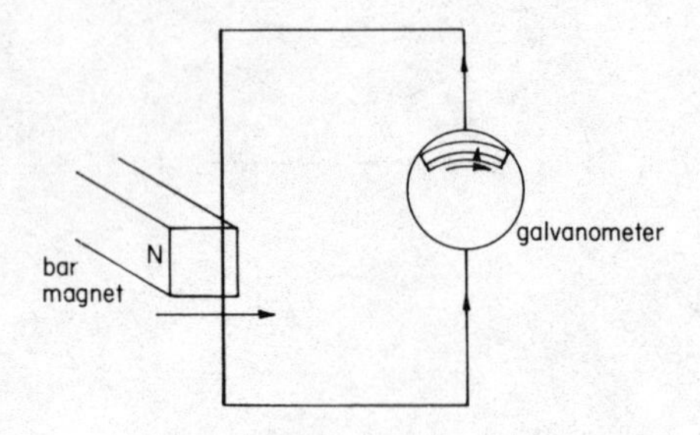

Figure 10.31

10.5.1 Faraday's Law of Induction

Faraday carried out tests on the induction effect and deduced that the magnitude of the e.m.f. produced was directly proportional to the *rate of flux cutting* by the conductor.

$$\text{Average rate of flux cutting} = \frac{\Delta\phi}{\Delta t}$$

where $\Delta\phi$ = quantity of flux cut in time Δt. The actual value of e.m.f. depends on the instantaneous value, which is denoted by $\mathrm{d}\phi/\mathrm{d}t$. We may understand this term $\mathrm{d}\phi/\mathrm{d}t$ better if we think in terms of the flux cut by a straight wire of length l as it moves along a path at right-angles to a field of flux density B (figure 10.32). Let the *steady* velocity of the wire be v m/s. Then area field cut per second is

$$A = l \times v \quad \mathrm{m^2/s}$$

And the quantity of flux enclosed by this area is

$$\phi = B \times A$$

$$= Blv \quad \mathrm{Wb}$$

This is the flux cut by the wire per second, or the rate of flux cutting.

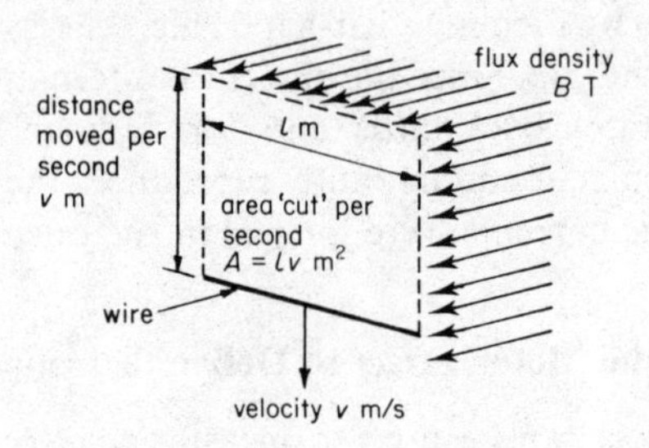

Figure 10.32

Hence for this simple case

$$\frac{\mathrm{d}\phi}{\mathrm{d}t} = Blv$$

We can then say that if the induced e.m.f. $E = \mathrm{d}\phi/\mathrm{d}t \times$ a constant, then

$$E = Blv \times \text{a constant}$$

If E is measured in volts, B in tesla, l in metres and v in metres per second the value of the constant is 1.0, thus the induced e.m.f. in volts is given by

$$E = Blv$$

We can prove this if we consider the quantity of work energy converted into electrical energy during one second. The work is done by the force required to propel the conductor across the field and out of the flux.

$$\begin{aligned}\text{Work done per second} &= F \times \text{distance moved per second}\\ &= Fv \quad \mathrm{J}\end{aligned}$$

But $F = BIl$ N (reverse of motor effect), therefore

$$\text{work done} = BIlv \ \mathrm{J}$$

Now electrical energy per second $= EI$ J (see section 9.5), therefore

$$EI = BIlv$$

or

$$E = Blv \text{ V}$$

Example 10.6

A straight conductor 0.7 m long is moved with a constant velocity of 20 m/s in a direction perpendicular to a uniform magnetic field. A voltmeter connected across the ends of the conductor shows a reading of 5 V. From these figures calculate the strength of the magnetic field.

Solution Let the strength of the field (flux density) $= B$ T. Then

$$\begin{aligned}\text{flux cut by conductor in each second} &= B \times \text{area crossed by conductor}\\ &= B(l \times v) \text{ Wb/s}\end{aligned}$$

But e.m.f. induced, E = rate of change of flux cutting by conductor, therefore $E = Blv$ V, hence

$$B = \frac{E}{lv}$$

$$= \frac{5}{0.7 \times 20}$$

$$= 0.357 \text{ T}$$

10.5.2 Direction of the Induced E.M.F.

So far we have only considered the magnitude of the induced e.m.f., but experiments show that it is produced in a direction which tends to oppose the agency from which it arises. For example, let the conductor form part of an external circuit in which the induced e.m.f. can cause a current to flow. The current passing through the conductor will then *create* a magnetic field whose flux opposes the flux of the original field cut by the conductor (figure 10.33). This is known as *Lenz's law of induction.*[6] For this reason the induced e.m.f. is referred to as a 'back e.m.f'.

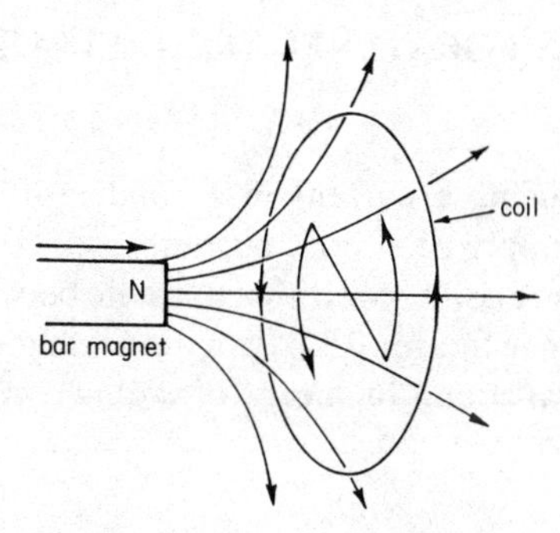

Figure 10.33 Induced current caused by moving magnet produces magnetic field in opposition to that of magnet

10.5.3 Fleming's Right-hand Rule

The relationship between the induced current direction, the direction of the *original* field and the direction of the motion of the conductor *with respect to the field* is indicated by the directions of the middle finger, forefinger and thumb of the *right* hand, disposed mutually at right-angles (figure 10.34).

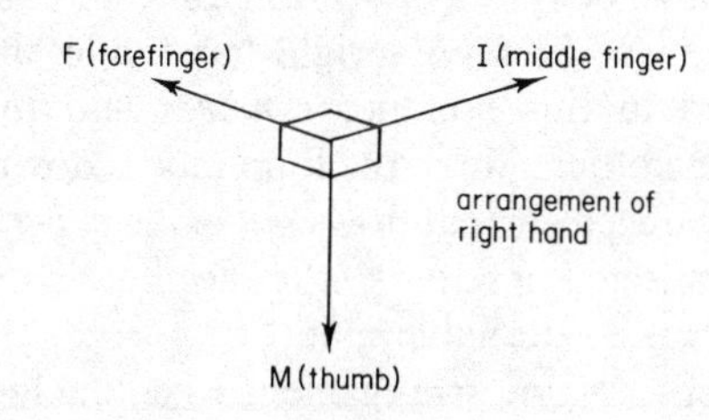

Figure 10.34 Fleming's right-hand rule

10.6 MACHINES FOR GENERATING ELECTRIC CURRENT

The effect of inducing a current in a conductor is used for the generation of a supply of electric current.

Consider a square coil or loop able to rotate between the poles of a magnet as shown in figure 10.35. Provision is made to connect the loop to an external circuit by means of sliding contacts called slip rings.

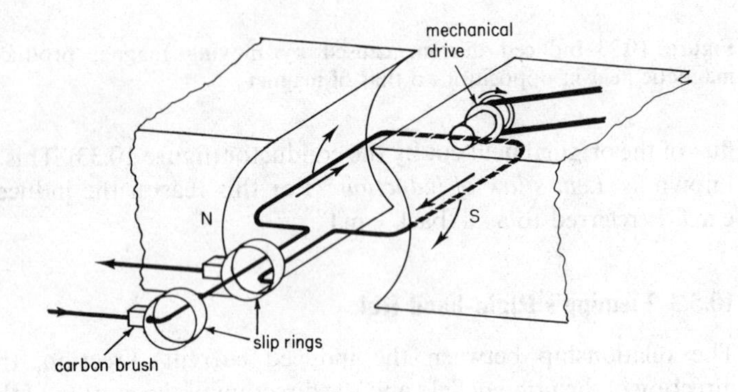

Figure 10.35 Principle of the alternator

As the loop rotates clockwise the left-hand conductor moves upwards, with a velocity v, at right-angles to the direction of the field. Application of Fleming's right-hand rule shows that the current induced in this conductor *moves into the paper*. The conductor on the other side of the loop moves downwards across the field so its induced current flows *out of the paper*. Both sides of the loop are connected in series so the effects are cumulative and a current flows in the external circuit.

Now consider the same arrangement when the loop has turned through an angle α (figure 10.36). The velocity of movement in a direction at right-angles to the direction of the field is now reduced and has the value $v \cos \alpha = v \sin \theta$, where θ is the angle measured from the vertical plane; so the magnitude of the induced e.m.f. is also reduced by the same proportion.

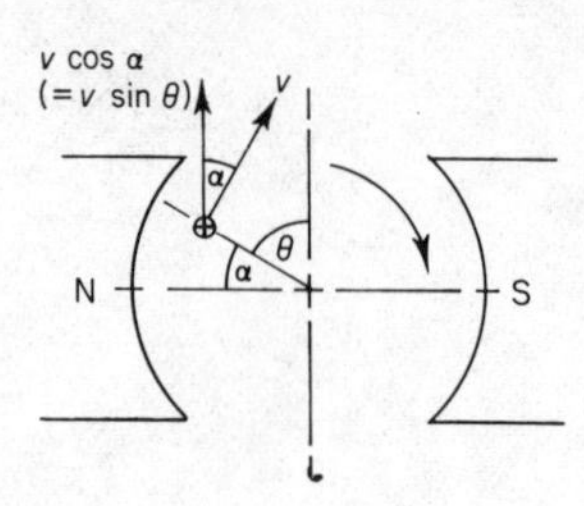

Figure 10.36

$$E = Blv_{\max} \sin \theta = E_{\max} \sin \theta$$

When $\theta = 0°$, $\sin \theta = 0$, hence the conductor is no longer cutting the flux of the field and the induced e.m.f. and current are zero. The change of current with the angle θ is shown in figure 10.37. As the conductors pass this position they start to cut the flux in the opposite sense as they re-enter the field and the direction of the induced current is then reversed (figure 10.38). For the full 360° of one revolution the current graph is as shown in figure 10.39. The current flows in opposite directions for each half of the complete *cycle*. In subsequent revolutions the process is repeated (figure 10.40). This is known as an *alternating current* (a.c.). The shape of the graph is said to be *sinusoidal*.

The number of cycles per second is the same as the rotational

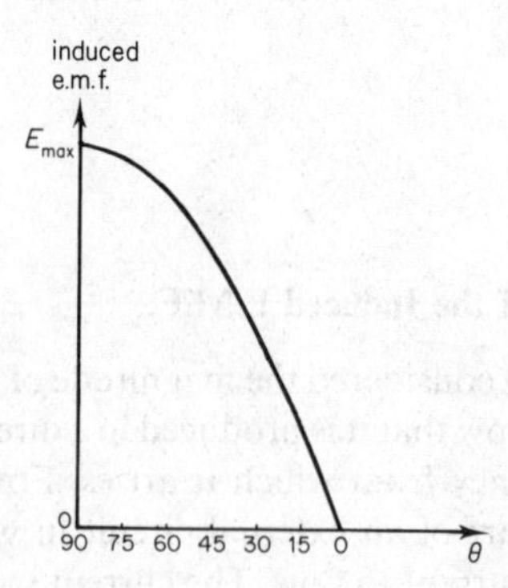

Figure 10.37

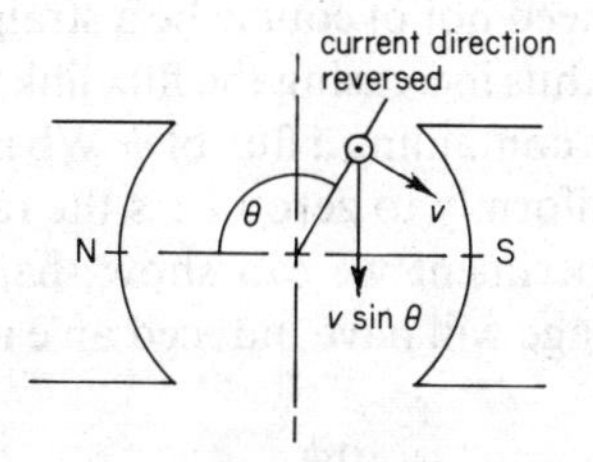

Figure 10.38

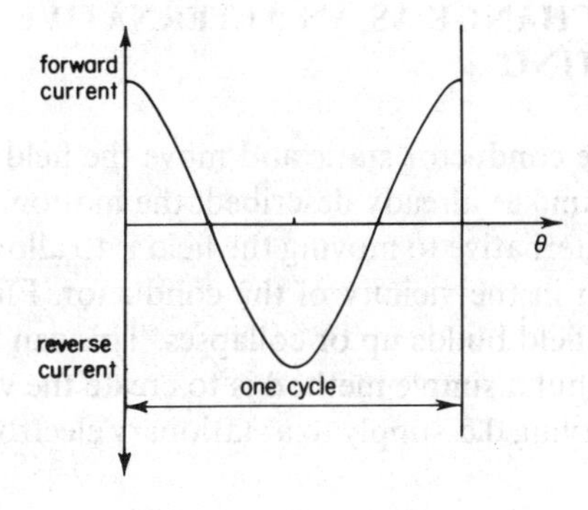

Figure 10.39

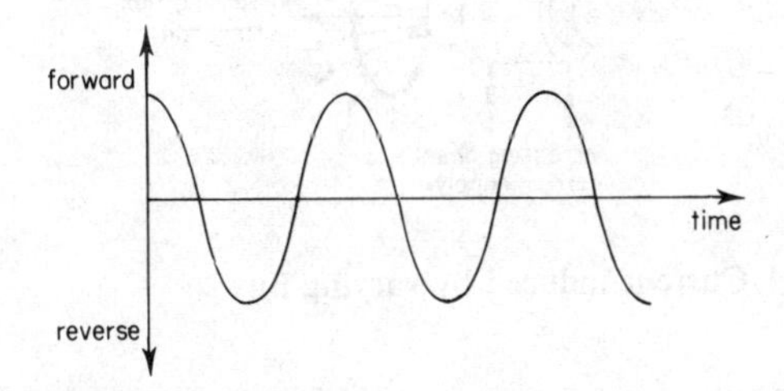

Figure 10.40

speed of this simple machine (or alternator) and is called the *frequency*. Its value is measured in units called *hertz* (Hz). British supply systems operate at 50 Hz, American systems at 60 Hz. The effective values of voltage and current from such a generator are discussed in section 9.7.

Example 10.7

A simple a.c. generator (alternator) consists of a flat coil with sides 0.1 m long rotating at a speed of 2000 rev/min in a uniform field of 0.6 T. The width of the coil is 0.06 m. Calculate (a) the frequency of the a.c. produced, in Hz, (b) the maximum speed of the conductors across the field, (c) the maximum (peak) voltage induced in the windings.

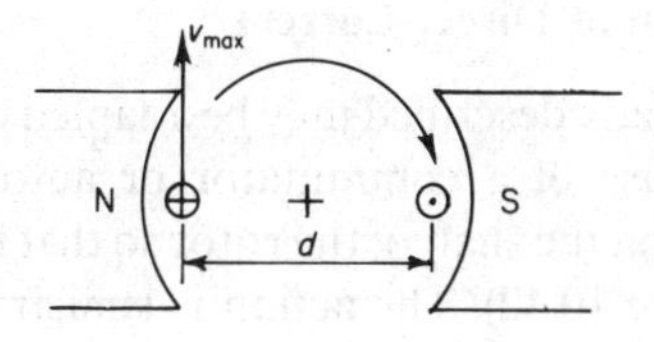

Figure 10.41

Solution (See figure 10.41) (a) The machine produces one full cycle of current per revolution, hence

frequency = number of full cycles per second
= number of revolutions per second

$$= \frac{2000}{60}$$

= 33.33 Hz

(b) Maximum speed of conductors across field is

$$\text{circumferential distance travelled per second} = \frac{2000}{60} \times \pi \times \text{diameter}$$

$$= \frac{2000}{60} \times \pi \times 0.06$$

$$= 6.284 \text{ m/s}$$

(c) Peak voltage is the e.m.f. induced when conductors are moving at maximum speed across field, thus

$$E = Blv$$
$$= 0.6 \times (2 \times 0.1) \times 6.284$$
$$= 0.754 \text{ V}$$

10.6.1 Generation of Direct Current

The machine already described may be adapted to produce a direct current by the use of a commutator or automatic changeover switch mounted on the shaft of the rotor so that it operates twice in every cycle (figure 10.42). The action is similar to that of the d.c. motor already described. The current graph will then be as shown in figure 10.43, that is, the current, although fluctuating in value, always flows in the same direction.

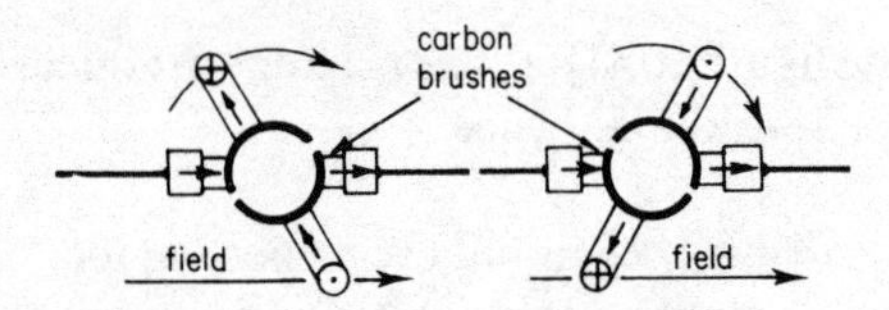

Figure 10.42 Use of commutator for direct current

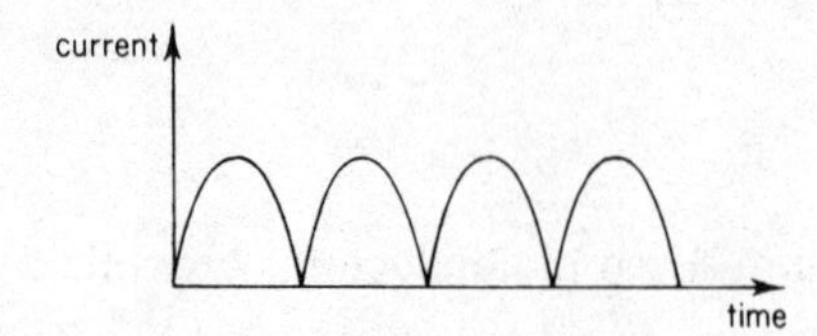

Figure 10.43

The conductor need not of course be a straight wire and is often wound into a coil, thus increasing the flux linkage. The flux linkage for a coil of N turns containing a flux of ϕ Wb is $N\phi$ Wb. If a flux of 1 Wb collapses uniformly to zero in 1 s the rate of change of flux $d\phi/dt = 1$. By experiment we can show that a circuit with unit change in flux linkage will have induced an e.m.f. of 1 V, therefore

$$\text{e.m.f. induced, } E = -N\frac{d\phi}{dt}$$

10.7 FLUX CHANGE AS AN ALTERNATIVE TO FLUX CUTTING

If we keep the conductor static and move the field, the induction effect is the same as already described; the motion only has to be *relative*. An alternative to moving the field is to allow an alteration in its strength in the vicinity of the conductor. Flux cutting still occurs as the field builds up or collapses. This can be arranged in several ways, but a simple method is to create the varying field by means of a varying d.c. supply to a stationary electromagnet (figure 10.44).

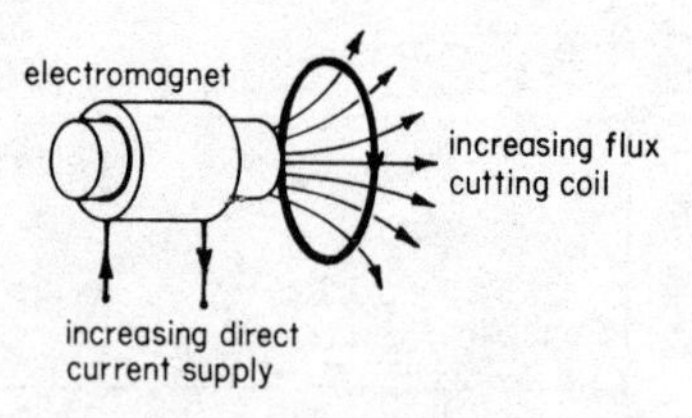

Figure 10.44 Current induced by varying flux

The effect is enhanced if the conductor in which the induction occurs is wound in a coil. When the current in the electromagnet is increased, the strength of the field increases and more flux cuts the static conductors of the coil. This effect is called a *flux linkage* because the flux is linked with the coil. As a result of the flux cutting the conductors an e.m.f. is induced in the coil. A very abrupt change of flux (a large $d\phi/dt$ value) occurs when the electromagnet is switched on or off, and it is during these operations that the induced e.m.f. is large.

Example 10.8

A coil of 10 turns is subject to a flux change from zero to 0.15 Wb in 2 s. Calculate the e.m.f. induced between the ends of the coil connections.

Solution

$$\text{Rate of change of flux } \frac{d\phi}{dt} = \frac{\text{change of flux}}{\text{time taken}}$$

$$= \frac{0.15 - 0}{2}$$

$$= 0.075 \text{ Wb/s}$$

The rate of change of flux is experienced by *every turn* of the coil. Therefore total flux change associated with the coil is

$$\frac{d\phi}{dt}_{\text{total}} = 0.075 \times \text{number of turns}$$

$$= 0.075 \times 20$$

$$= 1.5 \text{ Wb/s}$$

Hence

$$\text{induced e.m.f., } E = \text{total rate of change of flux}$$

$$= 1.5 \text{ V}$$

10.7.1 Applications of Induction by Flux Linkage

Car Coil Ignition System

This consists of two coils wound over the same iron core. One coil

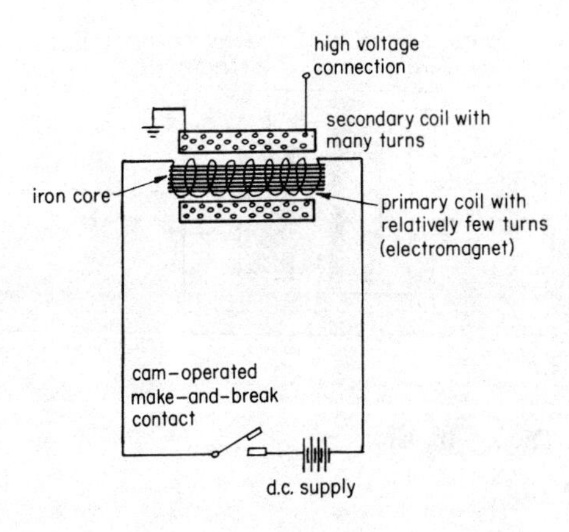

Figure 10.45 Coil ignition system

of thicker wire with relatively few turns is connected to a low voltage supply via a 'make-and-break' interrupter switch or contact breaker (figure 10.45). This coil acts as a source of flux. The second coil is wound on the same axis as the first so there is a large amount of flux linkage with its windings. When the contact is 'made' the current in the first rapidly increases to a maximum and the flux grows correspondingly with a high $d\phi/dt$ value. This flux linkage induces a very large e.m.f. in the second coil because the flux is linked with a very large number of turns of wire in the windings. When the contact 'breaks' the current rapidly decays in the low voltage coil, the field collapses causing a large negative value of $d\phi/dt$, and an induced e.m.f. is *again* created in the high voltage winding. These high voltage pulses only exist while the flux is changing in value; they are then used to operate the spark plugs of the engine.

The Transformer

This also uses two separate coils, one to generate the flux and the other to produce the induced e.m.f. The flux linkage is provided by winding the two coils on a common iron core of special design (figure 10.46).

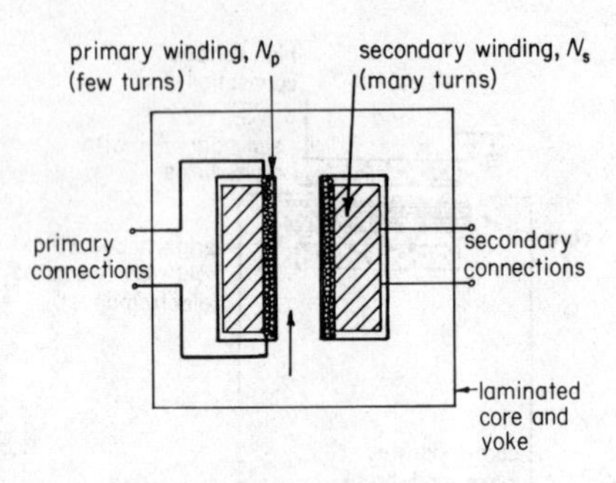

Figure 10.46 The a.c. transformer

The input winding (called the *primary*) is supplied with an alternating current (see section 10.6) which has a continuously varying magnitude. This causes continuous variations of a similar character in the flux. This varying flux is linked with the output (*secondary*) winding and induces an e.m.f. which also alternates in magnitude and direction.

The magnitude of the output e.m.f. is related to the supply voltage as follows. Let the variable flux produced in the core have a rate of change $\mathrm{d}\phi/\mathrm{d}t$, then the e.m.f. in the induced secondary winding is

$$E_s = \frac{N_s \mathrm{d}\phi}{\mathrm{d}t}$$

hence

$$\frac{\mathrm{d}\phi}{\mathrm{d}t} = \frac{E_s}{N_s}$$

Now the primary winding is also linked with the same flux so a back e.m.f. is induced in it (opposed to the supply voltage) having a value

$$E_p = \frac{N_p \mathrm{d}\phi}{\mathrm{d}t}$$

hence

$$\frac{\mathrm{d}\phi}{\mathrm{d}t} = \frac{E_p}{N_p}$$

But $\mathrm{d}\phi/\mathrm{d}t$ is the same for both windings so we may say

$$\frac{E_p}{E_s} = \frac{N_p}{N_s}$$

The magnitude of the back e.m.f. E_p is approximately equal to the supply voltage, if we ignore the losses involved in overcoming the resistance of the primary. Therefore

$$\frac{\text{output voltage}}{\text{input voltage}} = \frac{\text{output turns}}{\text{input turns}}$$

(turns ratio). Note that without a varying input current there will be no flux change ($\mathrm{d}\phi/\mathrm{d}t = 0$) and no induced e.m.f. in the output. Thus the transformer is only used with *alternating currents*. It is a most useful device because it can raise or lower voltages in a.c. equipment and has a very high efficiency.

The transmission of currents over long distances is possible if high voltages are used and this is made possible by transformers installed at either end of the transmission line (figure 10.47). Other applications include the provision of low voltage supplies for domestic appliances (such as door bells) or for situations where mains voltage would be dangerous.

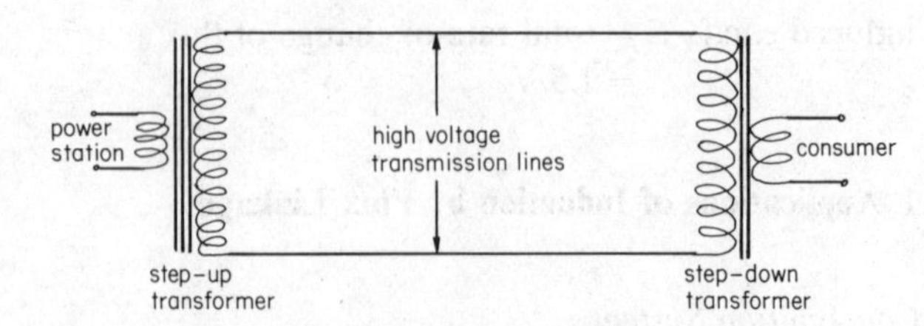

Figure 10.47 High voltage transmission of power

Example 10.9

An a.c. supply has a voltage of 240 V and is to be used for a bench system at about 20 V. Transformers are available with turns ratios of 20:1, 16:1, 5.5:1, 11:1 and 13:1. Which would be most suitable for this purpose?
Solution

$$\text{Ideal turns ratio} = \frac{\text{input voltage}}{\text{output voltage}}$$

$$= \frac{240}{20}$$

$$= 12:1$$

Try 11:1

$$\text{output volts} = \frac{240}{11}$$

$$= 21.82 \text{ V}$$

$$\text{error} = \frac{1.82}{20} \times 100$$

$$= 9.1\%$$

Try 13:1

$$\text{output volts} = \frac{240}{13}$$

$$= 18.46 \text{ V}$$

$$\text{error} = \frac{1.54}{20} \times 100$$

$$= 7.7\%$$

Hence nearest suitable ratio is 13:1.

10.8 SELF-INDUCTANCE AND MUTUAL INDUCTANCE

This effect, discovered by Henry[7] in 1831 relates to electromagnetic induction, in which a conductor experiences an induced e.m.f. when it is subject to a changing flux.

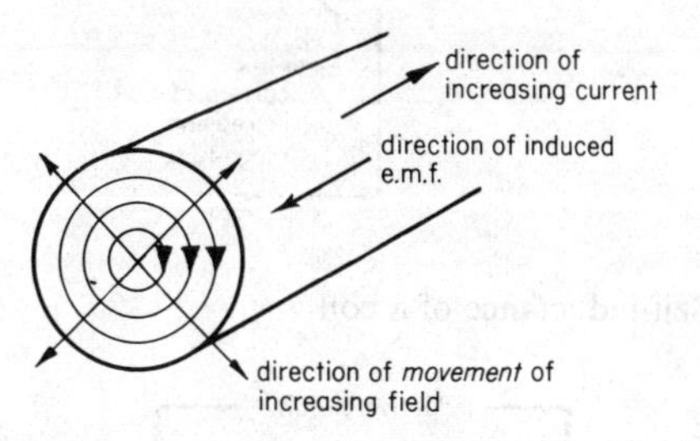

Figure 10.48 Self-inductance

If a conductor carries a current, it sets up its own concentric magnetic field and if the current is varying the flux will also be varying. The conductor is thus subject to a varying flux originating within itself; consequently an induced e.m.f. is generated in the conductor (figure 10.48). Inspection of the directions of flux change and field show that the induced e.m.f. will be in opposition to the direction of the *rate of change* of current. (*Note* not necessarily the direction of the current.) If the current is increasing, the induced e.m.f. will *oppose* the current, and if the current is decreasing the e.m.f. will be *with* the current. If the conductor is in the form of a coil the effect is perhaps easier to understand since the direction of the flux is defined (figure 10.49).

The effect is called *self-inductance*. If the change in current is very great, such as follows the opening of a switch, the induced e.m.f. may be large enough to drive a current across the air gap of the switch in the form of a spark. This is undesirable in switches because the contacts are damaged, so a capacitor is often used to absorb the energy of the system (figure 10.50) by allowing the e.m.f. to charge the capacitor rather than produce a spark (known as spark 'quenching').

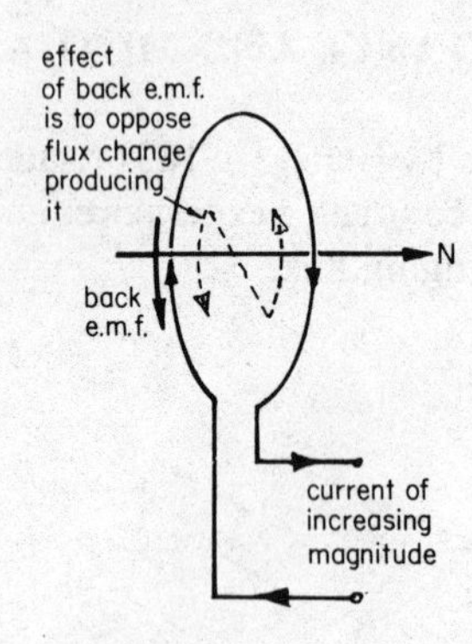

Figure 10.49 Self-inductance of a coil

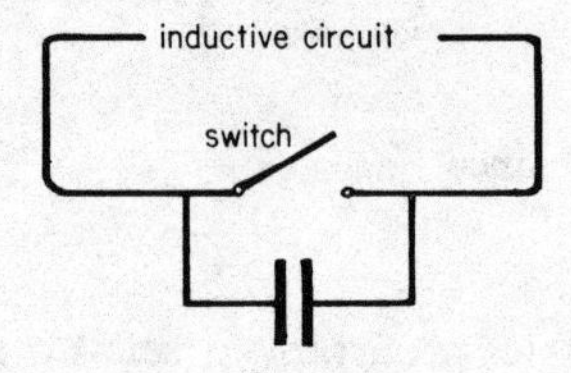

Figure 10.50 'Spark quencher'

The degree of self-induction depends on

(1) the rate of change of current $\mathrm{d}i/\mathrm{d}t$ and hence flux $\mathrm{d}\phi/\mathrm{d}t$
(2) the shape of the conductor (coil, straight wire, etc.)
(3) the medium in which the flux is created (air, iron, etc.).

For a given device (2) and (3) are constant quantities so we can say the induced e.m.f. is given by

$$E \propto \frac{\mathrm{d}i}{\mathrm{d}t}$$

$$E = \frac{\mathrm{d}i}{\mathrm{d}t} \times \text{a constant}$$

This constant is called the *inductance* of the device and has the symbol L, thus

$$E = L\frac{\mathrm{d}i}{\mathrm{d}t}$$

The units of L are henrys (H). The henry is defined as the self-inductance of a conductor in which a current changing at a rate of 1 A/s produces an e.m.f. of 1 V across the ends of the conductor.

Example 10.10

Calculate the self-inductance of a coil in which a back e.m.f. of 1000 V is detected when the current flowing changes uniformly from 2 A to 8 A in 0.7 s.

Solution Rate of change of current is

$$\frac{\mathrm{d}i}{\mathrm{d}t} = \frac{8-2}{0.7}$$

$$= 8.57 \text{ A}$$

$$\text{back e.m.f.} = L\frac{\mathrm{d}i}{\mathrm{d}t}$$

hence

$$\text{self-inductance, } L = \frac{E}{(\mathrm{d}i/\mathrm{d}t)}$$

$$= \frac{1000}{8.57}$$

$$= 116.7 \text{ H}$$

10.8.1 Mutual Inductance

There are many circumstances where a flux change in one coil also affects a neighbouring coil, as in the transformer and coil ignition systems (see section 10.7). An e.m.f. is induced in the *secondary* due to a *variation in current* in the *primary*. Therefore we may relate these two quantities by their mutual inductance M, where

$$E_s = M\left(\frac{di}{dt}\right)_p$$

M is also measured in henrys and is defined in a similar way to self-inductance.

Example 10.11

Two coils wound on the same former have a mutual inductance of 70 mH. An e.m.f. of 20 V is detected in the secondary coil when the current in the primary changes uniformly from zero to 10 A. Calculate the time taken for this change to occur.

Solution

$$E_s = M\left(\frac{di}{dt}\right)_p$$

hence

$$\left(\frac{di}{dt}\right)_p = \frac{E_s}{M}$$

$$= \frac{20}{70 \times 10^{-3}}$$

$$= 285.7 \text{ A/s}$$

Now di/dt = change of current/time taken, hence

$$\text{time taken} = \frac{10 - 0}{285.7}$$

$$= 0.035 \text{ s}$$

10.9 PRODUCTION OF MAGNETISM IN MAGNETIC MATERIALS

The causes of magnetism have been investigated closely and are now associated with the force exerted by a *moving* electric charge. (This is distinct from the forces exerted by *static* electric charges which have a different behaviour.) These forces are recognisable wherever an electric current is flowing (see section 10.2), but this does not readily explain the phenomenon of magnetic materials in which no current is flowing.

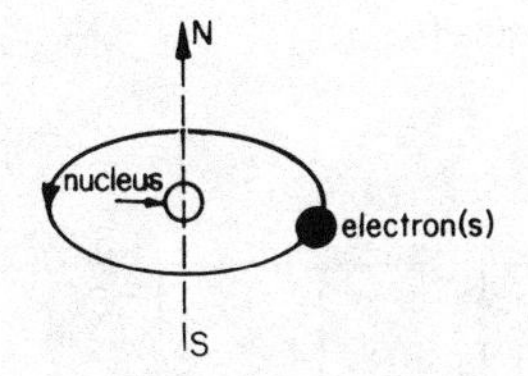

Figure 10.51 Automatic electromagnet

Atoms of iron and similar materials each possess a number of electrons which gyrate about the nucleus of the atom and create a tiny magnetic field at right-angles to their plane of rotation (as if the electron were flowing through a coil)—see figure 10.51. In an unmagnetised bar of iron these atomic magnets are disposed in small areas called *domains* with all their fields parallel, but different domains are aligned in different directions and cancel out (figure 10.52). If an *external field* is imposed on the material those domains aligned in the direction of the field grow larger and the remainder become smaller. As a result a stronger magnetic effect is produced (figure 10.53). Permanent magnet materials such as steel can retain the new configuration indefinitely but other materials such as soft iron quickly revert to their original dispositions and the over-all magnetic effect disappears.

10.9.1 Magnetising Force and Hysteresis

When magnetism is produced in a material by placing it in the field

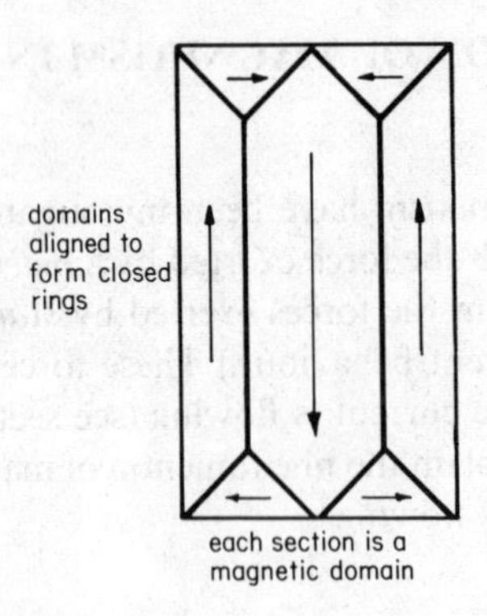

Figure 10.52 Unmagnetised state

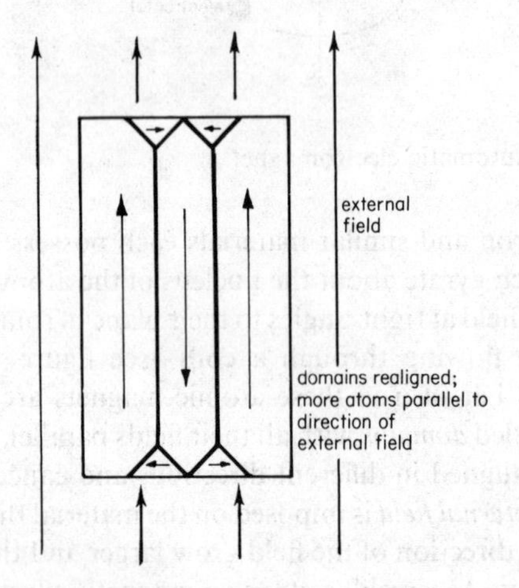

Figure 10.53 Magnetised state

of an electromagnet, the magnetic effect produced depends on both the *flux density*, *B*, produced in the material, and the *permeability* of the material (see section 10.3). The combination of these two quantities is called the *magnetising force*, *H*, where

$$H = \frac{B}{\mu}$$

(H is also known as the magnetising intensity and the magnetic intensity.) It can be shown that the units of H are ampere-metres since it is related to the magnetising current, and we may note that H is independent of the quality of the material being magnetised.

When an increasing magnetising force is applied to a material (for example, soft iron) the flux density, B, produced in the material, also increases. The relationship is not linear at first, but after an initial curve the graph of B against H straightens out and becomes proportional. The initial stage involves energy being absorbed by 'elastic' movement of the domain 'walls' in the material which does not appreciably increase the magnetic effect. If H is removed at this stage B returns to zero (figure 10.54).

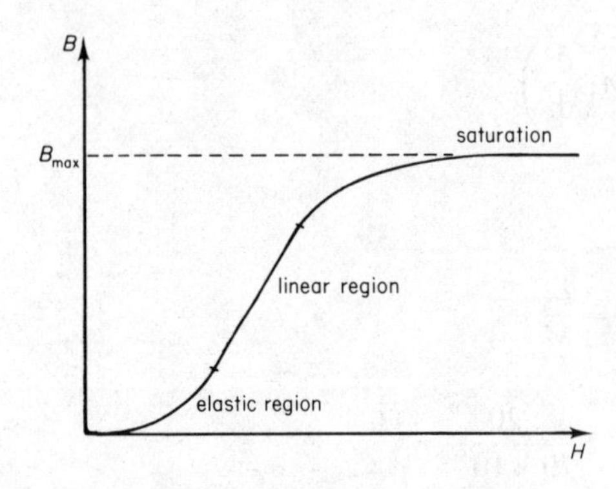

Figure 10.54 Initial magnetisation curve

Eventually the value of B reaches a maximum value called *saturation* and further increases in H have no effect. When the magnetising force H is reduced the flux density B also diminishes, but at a slower rate than we would expect. In fact when H is back to zero the value of B is still appreciable. This is called *remanence* and is due to the retention of the magnetised configuration (figure 10.55).

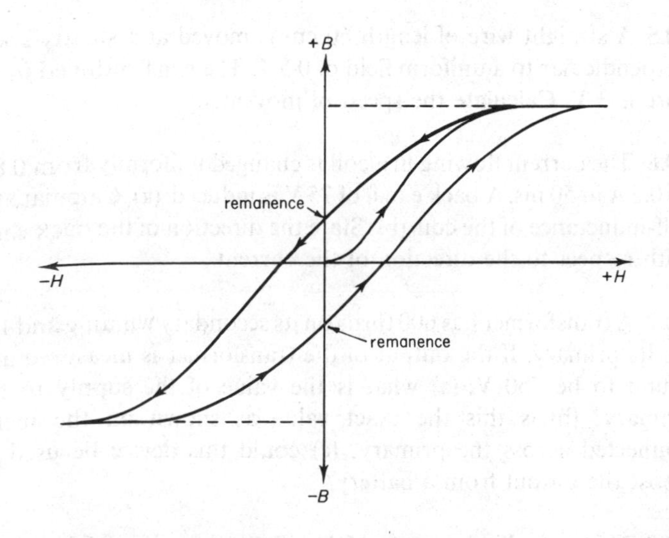

Figure 10.55 Hysteresis loop

If the magnetising current is now reversed and therefore the direction of H is reversed

(1) the value of B falls progressively to zero

(2) B increases in the opposite direction until saturation is reached.

If H is again reduced to zero and then imposed in the forward direction B follows a similar pattern, showing remanence and eventual saturation in the forward direction.

The forward and reverse graphs are separated by a 'loop' which represents a quantity of energy dissipated by the process. The manner in which B lags behind H is called *hysteresis*, meaning 'delayed', and the loop is a *hysteresis loop*, representing energy known as a *hysteresis loss*. The energy appears as heat in the material. In large magnetic devices handling alternating currents (for example, transformers) the quantity of heat evolved can be large and arrangements must be made for cooling by circulation of air or oil. The pipes used for this purpose are visible on the outside of large transformers such as those used in the electricity supply system.

1 Michael Faraday (1791–1867) was an English scientist. He devised the idea of 'lines of force' to construct images of magnetic flux (he was unable to do it mathematically); the idea was developed more formally by James Maxwell 50 years later. Faraday was also distinguished for his work on electrolysis and electromagnetic induction; he invented the electric generator.

2 Hans Oersted (1777–1851) was a Danish physicist. He discovered the magnetic effect of electric current (1819) in an experiment with compass needles placed over a conductor. His foundations of the study of electromagnetism were developed by Faraday and Henry.

3 Wilhelm Weber (1804–91) was a German scientist. He carried out work on magnetism with K. F. Gauss who had devised a logical system of units for magnetism; Weber devised a similar system for electricity in 1846.

4 Nikola Tesla (1856–1943) was a Croatian-born American electrical engineer. He was involved in the development of the transformer and its use in a.c. distribution of power supplies. He devised the Tesla coil—a high voltage induction coil.

5 Sir John A. Fleming (1849–1945) was an English electrical engineer. He worked under Maxwell and devised the easy rules for interpreting the fundamental relationship between electrical, magnetic and mechanical forces; he pioneered work on thermionic valves.

6 Heinrich Lenz (1804–65) was a Russian physicist. He devised rules relating the direction of induced e.m.f. to its cause; he established that the resistance of a conductor changes with temperature.

7 Joseph Henry (1797–1878) was an American physicist. He developed electromagnets for many applications, including the relay, which was applied to the electric telegraph by Samuel Morse. Henry is chiefly remembered for his work on electromagnetic induction, but he also designed the first electric motor in 1831.

TO THE STUDENT

At the end of this chapter you should be able to

(1) recognise the special effects which are known by the term 'magnetism' and understand their origin
(2) understand the concept of a magnetic field and measure its strength in different circumstances
(3) recognise the magnetic effects of an electric current and the force produced on a conductor
(4) understand the principle of the motor and similar devices
(5) understand the principle of electromagnetic induction and measure its effect
(6) understand the basic design of generation equipment and other applications of induction such as the transformer
(7) understand the concept of inductance as a circuit value
(8) understand the effects of magnetism on magnetic materials and the problem of hysteresis
(9) complete exercises 10.1 to 10.8.

EXERCISES

10.1 Calculate the flux passing through a coil of diameter 4 cm if the flux density of the uniform field is 0.07 T.

10.2 Determine the current flowing in a cylindrical coil of 600 turns and 60 cm long if the flux density of the field along the centre of the coil is 0.15 T ($\mu_0 = 4 \times 10^{-7}$ H/m).

10.3 A suspended cable is 60 m long between supports and lies at right-angles to the Earth's magnetic field. A current of 60 A (d.c.) flows along the cable. Assuming the Earth's field strength is 18 μT, calculate the force exerted on the supports due to the flow of current.

10.4 Calculate the maximum torque produced by a rectangular coil of 60 turns, 6 cm by 15 cm, whose long side is perpendicular to a parallel field of 0.2 T when carrying a current of 4 A.

10.5 A straight wire of length 60 cm is moved at a steady speed perpendicular to a uniform field of 0.5 T. The e.m.f. induced in the wire is 3 V. Calculate the speed of movement.

10.6 The current flowing in a coil is changed uniformly from 0.8 A to 0.2 A in 50 ms. A back e.m.f. of 75 V is induced. (a) Calculate the self-inductance of the coil. (b) State the direction of the back e.m.f. with respect to the direction of the current.

10.7 A transformer has 600 turns on its secondary winding and 120 on its primary. If the output of the transformer is measured and found to be 750 V, (a) what is the value of the supply to the primary? (b) is this the exact value as shown on the meter connected across the primary? (c) could this device be used to boost the output from a battery?

10.8 The mutual inductance, M, between two coils is 0.75 H. Find the e.m.f. induced in one coil when the current through the other changes uniformly from zero to 3 A in 2.5 s.

NUMERICAL SOLUTIONS

10.1 8.8×10^{-5} Wb
10.2 375 A
10.3 0.065 N
10.4 0.432 N m
10.5 10 m/s
10.6 (a) 6.25 H
10.7 (a) 150 V
10.8 0.9 V